DUTY AND THE BEAST

The moral status of animals is a subject of controversy both within and beyond academic philosophy, especially regarding the question of whether and when it is ethical to eat meat. A commitment to animal rights and related notions of animal protection is often thought to entail a plant-based diet, but recent philosophical work challenges this view by arguing that, even if animals warrant a high degree of moral standing, we are permitted – or even obliged – to eat meat. Andy Lamey provides critical analysis of past and present dialogues surrounding animal rights, discussing topics including plant agriculture, animal cognition, and in vitro meat. He documents the trend toward a new kind of omnivorism that justifies meat-eating within a framework of animal protection, and evaluates for the first time which forms of this new omnivorism can be ethically justified, providing crucial guidance for philosophers as well as researchers in culture and agriculture.

ANDY LAMEY is Assistant Teaching Professor in the Department of Philosophy at the University of California, San Diego. He is the author of *Frontier Justice: The Global Refugee Crisis and What to Do about It* (2011).

DUTY AND THE BEAST

Should We Eat Meat in the Name of Animal Rights?

ANDY LAMEY

University of California, San Diego

CAMBRIDGE
UNIVERSITY PRESS

CAMBRIDGE
UNIVERSITY PRESS

University Printing House, Cambridge CB2 8BS, United Kingdom

One Liberty Plaza, 20th Floor, New York, NY 10006, USA

477 Williamstown Road, Port Melbourne, VIC 3207, Australia

314–321, 3rd Floor, Plot 3, Splendor Forum, Jasola District Centre, New Delhi – 110025, India

79 Anson Road, #06–04/06, Singapore 079906

Cambridge University Press is part of the University of Cambridge.

It furthers the University's mission by disseminating knowledge in the pursuit of education, learning, and research at the highest international levels of excellence.

www.cambridge.org
Information on this title: www.cambridge.org/9781107160071
DOI: 10.1017/9781316672693

First published 2019

Printed and bound in Great Britain by Clays Ltd, Elcograf S.p.A.

A catalogue record for this publication is available from the British Library.

Library of Congress Cataloging-in-Publication Data
NAMES: Lamey, Andy, author.
TITLE: Duty and the beast : should we eat meat in the name of animal / Andy Lamey, University of California, San Diego.
DESCRIPTION: New York : Cambridge University Press, 2019. | Includes bibliographical references and index.
IDENTIFIERS: LCCN 2019003152 | ISBN 9781107160071 (alk. paper)
SUBJECTS: LCSH: Meat – Moral and ethical aspects. | Animal welfare.
CLASSIFICATION: LCC TX371 .L36 2019 | DDC 641.3/6–dc23
LC record available at https://lccn.loc.gov/2019003152

ISBN 978-1-107-16007-1 Hardback

Contents

Acknowledgments

I incurred debts to several institutions in completing this book. Although it is not a revised version of my doctoral dissertation, previous versions of four chapters did first appear there. I am grateful to my supervisor at the University of Western Australia (UWA), Miri Albahari, and to my external examiners, Michael Allen Fox, Richard Keshen and Peter Singer for their constructive feedback. In completing my dissertation I received a grant from the Culture and Animals Foundation that allowed me to attend a conference on the work of Peter Singer held at Sydney University in 2007 at which I presented an ancestor version of Chapter 5. More recently in 2017 I benefitted from the generosity of the Humane Society of the United States by attending Camp Muse in Shin Pond, Maine, a writer's retreat that the society makes available to researchers working on animal issues.

Ideas in this book were previously presented to audiences in the philosophy departments at Sydney (October 2011) and Monash University (April 2013); the Third Annual Oxford Animal Ethics Summer School (July 2016); a workshop on Veganism and Beyond: Food Animals and Ethics (June 2017), at Queen's University; the 2018 Minding Animals Conference at the National Autonomous University of Mexico (January 2018); and the Moral and Political Philosophy Seminar (October 2017) at my home institution, the University of California, San Diego. At UCSD I benefitted in particular from the feedback of my departmental colleagues Dana Nelkin and Sam Rickless, whose work informs Chapter 4.

Several chapters draw on previously published articles. Chapter 3 uses revised material from "Food Fight! Davis versus Regan on the Ethics of Eating Beef," *Journal of Social Philosophy* 38 (2007), 331–48 and "Field Deaths in Plant Agriculture," *Journal of Agricultural and Environmental Ethics* (2018), 1–20, co-authored with Bob Fischer. Chapter 6 uses revised material from "Primitive Self-consciousness and Avian Cognition,"

The Monist 95 (2012), 486–510. Chapters 3 and 4, respectively, contain a few paragraphs from my review of *The Moral Complexities of Eating Meat*, published in *Between the Species* 20 (2017), 133–46, and "Making the Animals on the Plate Visible: Anglophone Celebrity Chef Cookbooks Ranked by Sentient Animal Deaths," *Food Ethics* (2018), 1–21, co-authored with Ike Sharpless.

Hilary Gaskin at Cambridge University Press expressed an early interest in the idea for this book that made its eventual publication possible. I am also grateful to Mathew Fulkerson, Valéry Giroux, Steven Luper, Josh May, Nathan Nobis, and Jeff Sebo for commenting on individual chapters. I owe a special debt to Josh Milburn for offering insightful comments on the entire manuscript. My gratitude to Bob Fischer is multifaceted. Not only did Bob co-author one of the papers that forms the basis for Chapter 3 but he also provided probing comments on six other chapters. Readers of Chapter 9 will recognize Bob as the co-author of one of the papers that I criticize there. Although we may disagree on the ethics of in vitro meat I am grateful to Bob for publishing such a well-argued article, one that I knew I had to grapple with as soon as I read it, and for discussing many of the ideas in this book with me.

I owe a more personal debt to my wife Kirsty Best and my daughters Lasha and Lucy, all of whom tolerated the many hours of solitude I spent working on this book. This book is for them.

Introduction: The New Animal Debate

The Rise of New Omnivorism

One of the traditional assumptions of the debate over the ethical status of animals has long been that someone who is committed to reducing animal harm should not eat meat. Peter Singer, Tom Regan, and other philosophers associated with the animal rights movement – which is better referred to as the animal protection movement, for reasons outlined below – have advocated not eating animals. On the other side of the debate, philosophers who have rejected animal protection have commonly argued that animals occupy a lower moral status than protectionists aver before going on to invoke this as a premise in arguments for meat-eating. Despite the disagreements between the two camps, both have traditionally taken it for granted that the philosophy of animal protection and the practice of avoiding meat rise and fall together.

This book is about a new position in the debate. This position, which I term *new omnivorism*, endorses animal protection as philosophy but goes on to defend eating animals. Some versions of new omnivorism argue that eating animals is permissible even if protection theory is true. Others go further and argue that protectionist arguments actually *oblige* us to eat animals. What all versions of new omnivorism have in common is that they are impossible to capture in the debate's traditional terms. In order to understand the different forms of new omnivorism, therefore, it is helpful to briefly chart their rise and note how they differ from traditional challenges to animal protection.

The modern philosophical debate over the moral status of animals began in the 1970s.[1] Singer, Regan and other philosophers argued for improved

[1] For philosophical debates regarding animals before the modern period, see Sorabji (1995) and Garrett (2000).

treatment of animals while disagreeing on such questions as which ethical theory is best. Despite their disagreements, the thinking of protectionism's pioneers converged at the level of dietary ethics. Singer devoted a chapter of *Animal Liberation* to "Becoming a Vegetarian," while Regan in *The Case for Animal Rights* included a section titled "Why Vegetarianism is Obligatory" (Singer 1975: 159–84; Regan 1984: 330).

In the 1980s and 90s protectionist arguments were subject to sustained criticism by philosophers such as R. G. Frey (1980, 1983), Peter Carruthers (1992) and P. T. Leahy (1991). While these first-wave critics mounted their challenges in the name of different philosophical theories, such as utilitarianism (Frey), contractualism (Carruthers), and Wittgensteinianism (Leahy), they shared the view that protectionism was misguided. They generally argued that the common sense of society regarding animals at the time of writing was by and large defensible when it was not already too sensitive to the moral claims of animals. Carruthers for example saw his project as taking place in the wake of a "recent explosion of interest in animal rights," a philosophical development he regarded as pernicious (1992: xi). "Just as Nero fiddled while Rome burned, many in the West agonise over the fate of seal pups and cormorants while human beings elsewhere starve or are enslaved" (xi). Carruthers and other early critics rejected protectionism at the level of both its philosophical claims and its action-guiding recommendations. It was common, for example, for their works to criticize core protectionist notions such as the idea that animals can have welfare interests or that species membership is morally irrelevant. Similarly, where protectionists challenged factory farming, hunting, and other practices, these practices were, with occasional exceptions, defended by their first-wave critics.

Around the turn of the century, first-wave critiques of animal protection began to give way to a different kind of response. This response sought not to challenge protectionism at the level of first principles, but to disassociate such principles from the conclusion that eating animals is impermissible. Even if the philosophical claims protectionists have argued for are correct, these new defenders of omnivorism maintained, it does not follow that we are obliged to stop eating animals. We are either permitted or obliged to continue doing so, depending on the particular version of new omnivorism being advanced. Such arguments now appear in a diversity of forms, but all have in common the fundamental fact that they reject the link between protection theory and meatless eating that was once considered axiomatic by both defenders and critics of animal protection.

The difference between first and second wave critics of animal protection illustrates a distinction political theorists have drawn between *rejectionist* and *immanent* social critics (Walzer 1987: 64).[2] A rejectionist challenges a given social order from an independent or external point of view, while an immanent critic judges a society wanting according to values that its members widely accept. First-wave critics of animal protection criticized a school of philosophy rather than a society, but their stance toward that philosophy was nonetheless rejectionist. Arguments for new omnivorism are immanent critiques insofar as they employ protectionist values as premises, but conclude that those very values call into question protectionism's traditional recommendations regarding diet. Just as individual social critics have at times combined rejectionist and immanent arguments in a single work of criticism, the same has been true of critics of animal protection (e.g. Leahy 1994: 201; Posner 2004). But at the level of the overall debate, first-wave rejectionist responses to animal protection have been noticeably joined by second-wave immanent critiques. The more one engages the animal ethics literature now, the more on encounters defenses of meat-eating that take this form.

Varieties of New Omnivorism

Steven Davis (2003) is a paradigmatic example of a second-wave critic. What makes Davis's argument noteworthy is that it takes as its point of departure Regan's own argument for animal rights. Davis's twist is to argue that a diet that contains free-range beef is more consistent with Regan's theory than the plant-based diet Regan has long advocated. Davis arrives at this conclusion by pointing out that mice and other field animals are killed during crop cultivation. He posits that more animals are actually killed in the production of a plant-based diet than in the production of a diet containing free-range beef. In Davis's hands, such empirical claims are combined with a canonical theory of animal protection to entail a dietary ethic that, contrary to what protection theorists have long argued, ranks a meatless diet second-best to one containing some meat. Davis's view, which I term *burger veganism*, has now inspired other critics to offer variations on his argument that invoke protectionist premises to justify eating not only free-range beef, but free-range meats made from sheep,

[2] Walzer refers to *disconnected* rather than *rejectionist* criticism, but theorists who employ his distinction generally employ the latter label (e.g. Beiner and Nedelsky 2001: 289; Ober 2011: 48–9).

goats, kangaroos, and other species (Schedler 2005, Bruckner 2016; cf. Archer 2011a).

A different form of new omnivorism finds inspiration in the work of Temple Grandin. Grandin, who is well known as a leading representative of people with autism, has also designed a method of slaughtering animals that is more humane than the factory farming that was the norm when Singer and Regan wrote their pioneering works. Grandin's slaughter system is designed to exploit animals' natural behaviors so that they move through slaughter facilities on their own accord. Cattle, for example, are sensitive to loud noises, bright lights, and shadows, so Grandin's abattoirs eliminate such distractions. This removes the need for a handler to intervene with a cattle prod, to the point that cattle and other animals will freely insert themselves in the apparatus that restrains them as they are killed. Such a system sees the vast majority of animals rendered unconscious moments before slaughter, which makes the process less painful.

Grandin's system has achieved popularity in part because it is a manifestation of the increasingly popular view that while we have moral obligations to animals, they require only that we minimize their suffering and do not prohibit killing. So long as killing is done painlessly, this view holds, taking the life of a food animal is permissible. But while Grandin's writings contain articulations of this idea, her primary contribution to the debate over the ethical treatment of animals is not as a philosopher but as an engineer. Her technology has featured in a prominent critique of Singer that views his philosophy of animal liberation as having less radical implications than Singer has argued for. On this reading, which has been advanced by Gary Francione and other critics, Singer's argument for animal liberation entails not veganism, but the replacement of traditional slaughter methods with Grandin's (Francione 2008). This interpretation of Singer argues that no basis for an objection to painlessly killing animals for food and other purposes can be found in his work. On this account, the widespread belief that Singer's theory is a vegan or even vegetarian philosophy is based on little more than historical coincidence, resulting from the fact that Singer developed his theory in the pre-Grandin age, when our dietary alternatives were factory farmed meat or no meat at all. Today, however, Grandin's system is increasingly common. Singer's theory, it is argued, must therefore be decoupled from any systematic connection to vegetarianism, let alone veganism. As burger veganism does in the case of Regan, this critique leaves Singer's starting premises intact but calls into question one of their most important action-guiding conclusions.

In addition to the work of Davis and Grandin, new omnivorism finds a third source of support in the work of so-called "plant neurobiologists." This controversial group of botanists holds that the signaling and related capabilities of plants are more sophisticated than has been traditionally recognized. Philosophers sympathetic to plant neurobiology have gone so far as to argue that plants are actually sentient. In this way their work employs a concept crucial to protection theory, which has long been concerned with elucidating the moral standing of sentient animals. The fact that animals are sentient and plants are not is a bedrock feature of the protectionist case for granting moral standing to animals and withholding it from plants. The plant neurobiology view, however, "casts doubts upon the utility of the traditional rigid division made between plants and animals" (Pelizzon and Gagliano 2015: 5). The empirical claims of plant neurobiology are now cited as grounds not to draw dietary or other distinctions between animals and plants. Plant-based diets on this view are drained of any moral superiority over omnivorous ones, again in a manner that does not dispute the protectionist case for the moral standing of animals.

Burger veganism, Grandin's system of humane slaughter, and the neurobiology school of botany are all products of the twenty-first century. But if the turn of our century marks the beginning of the new omnivorist era, one immanent argument has a much older history. This argument takes over from protectionism the idea that animal lives have value. But it invokes this value to defend eating them on the grounds that food animals are only brought into the world to be eaten. The logic of the larder, as this argument is called, views plant-based diets as morally impermissible insofar as their widespread adoption would result in a world that lacked the value represented by the lives of food animals. An influential version of the logic of the larder was formulated in the late nineteenth century, making it more venerable than the above-mentioned forms of new omnivorism. The logic of the larder, however, shares with other forms of new omnivorism the fact that it obliges meat eating in a manner that is consistent with protectionist premises about the value of animal lives. This may be why it has been the subject of renewed interest during the same period in which immanent arguments have achieved prominence (Hare 1999: 240; Scruton 2000: 100, 60; Schedler 2005: 502–3; Callicott 2016: 59, cf. Matheny and Chan 2005; Višak 2013: 129–33).[3]

[3] To my knowledge the first writer to mention the logic of the larder during the modern debate over the ethical status of animals was Robert Nozick, albeit in the context of defending a traditional conception of animal rights and vegetarianism rather than new omnivorism (1974: 38).

A final challenge to protectionist dietary codes is represented by another product of the twenty-first century: in vitro meat. It is created by taking a cell from an animal and growing it in a laboratory into edible flesh. In vitro meat is thus identical to the real thing but for the fact that it is not carved out of the carcass of an animal. Current production methods involve the use of fetal bovine serum, a growth hormone taken from the fetuses of cows that are pregnant at slaughter. In this way, in vitro meat continues to involve harm to animals. Scientists involved in the creation of in vitro meat, however, are already working to develop plant-based alternatives. Given the realistic possibility of such a development, the long-term future of in vitro meat raises the possibility of a new form of meat that involves no harm to animals. As such, it represents yet another form of meat-eating that is arguably consistent with arguments for animal protection.

Assessing New Omnivorism

The purpose of this book is twofold. The first is to highlight how the animal debate has changed. In arguing with old-fashioned rejectionist critics, protectionists could simply make the case for their first principles. In protectionism's early days, such arguments were innovative, original, mind-expanding. Even though we live in a society that still assigns animals a lower moral status than protectionism argues for, in the context of the debate among philosophers, simply making the basic case for a higher status is no longer bold or innovative. Indeed, the philosophical literature now sees variations on protectionism advanced within the framework of equal consideration, rights theory, utilitarianism, contractualism, virtue ethics, and Kantianism – every prominent moral philosophy now has been applied to animal ethics. There is some value in having so many diverse arguments for protectionism, even if their conclusions sometimes seem repetitive. But it no longer suffices to make a basic case for protectionism. Something also needs to be said in reply to the critics who use traditional protectionist arguments against traditional protectionist conclusions. Their arguments are the philosophical equivalent of jiu-jitsu, which uses an opponent's own energy against her, and so represent an approach that cannot be countered with sheer assertiveness.

Providing the necessary response to new omnivorism is my second purpose for writing. My overall verdict is mostly negative. Burger veganism, I argue, rests on mistaken empirical calculations and an implausible refusal to distinguish unintentional from intentional harms. In arguing against both Davis's view and that of his philosophical descendants, I make a positive case for a new version of the doctrine of double effect (DDE) that, like previous versions, ranks harmful direct agency as worse than harmful indirect agency, but unlike earlier versions extends the DDE to harm directed at animals. Taking intentions into account in this way should cause us to see how Davis-style new omnivores err in ranking the death of one field animal just as wrong as that of a cow or other food animal killed deliberately. By employing the animal-friendly DDE, we can see that all else being equal, intentionally killing the cow is worse. An especially strong version of burger veganism makes the case for hunting and eating wild animals. Although this argument is not refuted by appealing to the animal-friendly DDE alone, my version of the DDE does sap much of its strength, and the argument ultimately fails on broader protectionist grounds.

Against the view that Singer-style protectionism permits humane slaughter, I note the many grounds on which a proponent of Singer's arguments can reject such a view. I refer to Singer's arguments in the plural because he has made two influential arguments regarding animals, one of which appeals to utilitarianism, the other to equal consideration, which is not an exclusively utilitarian value. Grandin's otherwise progressive system still permits systematic animal suffering, which both of Singer's arguments condemn. Singer's stronger, non-utilitarian argument would justify a system of humane slaughter in which the system really was painless and all the animals were merely sentient, and so lacked traits fundamental to personhood. But Singer's non-utilitarian argument ultimately fails to justify Grandin's system, for reason beyond the fact that the no-pain condition is clearly not met. Among other reasons, Singer's equal consideration argument fails to recognize that death harms even merely sentient beings by frustrating their interest in continued existence. A further problem for attempts to justify humane slaughter is that recent research suggests that chickens, the animals killed in by far the greatest number under Grandin's system, possess a rudimentary form of self-consciousness. This renders it morally wrong to kill them for food when non-harmful alternatives are widely available, not only according to Singer's equal-consideration account, but any theory that grants moral weight to the distinction between merely sentient beings and persons.

Arguments from plant neurobiology fail to show that plants are sentient as the term is normally understood. Insofar as philosophers sympathetic to plant neurobiology deploy a revised and more expansive notion of sentience, they fail to challenge the moral hierarchy that ranks animals over plants. Regarding the logic of the larder, I argue that it rests on a false understanding of good choices. A necessary condition of a choice being morally good (i.e., morally obligatory or supererogatory) is that failing to bring it about will be bad for someone. Insofar as failing to bring food animals into existence will not be bad for any individual, there is no failure to execute a morally good choice. There is a variation of the larder argument that would justify consuming a certain kind of genetically modified animal. The animals would be engineered to grow to full size and then painlessly die, so that they would not need to be killed. While consuming such animals would be permissible on my account, it is currently not scientifically possible to create them. If or when such a form of omnivorism does become possible, it would need to meet several conditions to be justified. It would be speciesist, for example, to permit such genetically modified food animals but not permit the creation of similarly modified human beings, whom we might use for such purposes as harvesting organs for transplant. Such scientific and ethical hurdles suggest that while science-fiction omnivorism, as I term it, would potentially be permissible it is also likely to remain improbable.

The one form of new omnivorism we should straightforwardly accept is that which condones eating in vitro meat. In my final chapter, I change from new omnivorism's critic to its defender by arguing in favor of eating such meat. I rebut protectionists who have criticized it for perpetuating a framework of intelligibility according to which animals are edible, among other perceived shortcomings. Against such critics, I argue that the arrival of in vitro meat should prompt us to reconceive meat as a substance that does not necessarily require killing animals and so does not harm them. In this way, meat eating ultimately can be shown to be safe for protectionism, marking an important development regarding the advocacy claims of animal protection.

Project Goals and Scope

In taking up this project, my goal is not to offer a stiff-necked defense of veganism, come what may. At different times in conceiving and writing this book, my conclusion as to what diet animal protection entails has changed. Indeed, in one sense this book represents my own attempt to sever the link between protection and meat-free diets. For, as I noted above, in vitro meat is a kind of meat protectionists should welcome, thereby ending the historic link between animal protection and plant-based diets. But in other ways this book represents a criticism of the critics, arguing that the mass slaughter of animals for food remains unjustified. In challenging new omnivorism I seek to defend the traditional view that endorsing animal protection entails a diet free of animal flesh. Against the widespread view of animal protection as a radical cause, therefore, this is a work of animal protection that adopts the stance of counter-revolution, arguing against reformers who threaten to upturn a pre-existing moral universe. But just as conservatives have long endorsed change to preserve a cherished order, so do I go beyond what has been said and offer new arguments in defense of protection theory's traditional dietary recommendations. They include a new formulation of the doctrine of double effect, a new empirically informed account of the moral standing of chickens – the most widely consumed land animal – and a new theory of procreative ethics.

Given these goals, this book spends only two chapters making a case for protectionism. These preliminary chapters outline what I believe to be the strongest version of protection theory. Rather than mount a case within the framework of a full-blown moral theory such as utilitarianism or deontology, which would have limited force with someone who was not a utilitarian or a deontologist, I appeal instead to mid-level moral principles that are in principle compatible with more than one philosophy. These include the equal consideration of interests, anti-speciesism and time-relative interests (and, in later chapters, the doctrine of double effect and primitive self-consciousness). To be sure, these principles require their own defenses, and they are not compatible with every last theory. Yet each can in principle be endorsed by a range of moral philosophies rather than just one. In this way, they direct our attention more closely to issues concerning the moral status of animals themselves, rather than familiar debates about the ultimate grounds of ethics, which show no sign of ending any time soon.

Most of my attention focuses directly on the arguments and technologies that are invoked to defend the diverse forms of new omnivorism. As we will see, this focus often requires countering specific philosophical arguments with a positive case for protectionism. If many of protectionism's leading critics now offer immanent critiques, immanence nonetheless remains a matter of degree. It is a mistake to view the debate over new omnivorism occurring between two sides that agree on philosophical principles and just happen to draw on different empirical or technological considerations.

A better image is conjured by Ronald Dworkin's notion of contemporary political philosophy taking place on an egalitarian plateau: liberals, libertarians, communitarians, feminists, analytic Marxists and the rest generally affirm some notion of equality, even as they diverge on what it entails (Dworkin 1983). Similarly, the debate about animals this book engages takes place on a broadly protectionist plateau. This, however, no more entails that there are no genuinely philosophical differences at stake than Dworkin's image entails that there are no philosophical disagreements between liberals, libertarians, and the rest.

Out of the many possible protectionist recommendations immanent critics might challenge, I focus on meat-eating because it is the practice that harms the most animals. Where clothing, entertainment, and medical science harm millions of animals, billions are killed for food. Animal protection theory has for this reason long been closely concerned with dietary questions. Taking up dietary debate inevitably requires examining empirical considerations. A widely held view among animal ethicists, for example, holds that the mental capabilities of animals matter for their moral status. Much hangs on whether or not a member of any given species (including ours) is self-conscious rather than merely conscious, a question contemporary science increasingly has bearing on. Similarly, it matters how many field animals are killed in plant cultivation and what inferences about plant sentience are and are not warranted by the claims of plant neurobiologists. Hence the overall approach of this book: applied ethics in the trenches, in which empirical considerations and moral principles both receive scrutiny and examination.

To note the relevance of empirical considerations is not to take them as a substitute for philosophical concerns. New omnivorists appeal to a diverse range of ethical arguments and concepts in making their case. These include not only familiar utilitarian and rights-based arguments for animal protection, but also theories of personhood and well-being, and the distinction (or lack thereof) between acts and omissions. A central issue in the debate concerns which side has a better formulation of these and other notions. The ways in which immanent critiques of animal protection could potentially be mounted are vast. Focusing on those that have bearing on dietary ethics thus imposes an order on my discussion. And as we will see, discovering what protectionism entails in this area is already difficult enough.

If I do not assess immanent arguments on topics beyond diet, there are also broadly immanent arguments for meat eating that I set aside. One is the argument from the so-called inefficacy objection. It holds that however sound the case for animal protection may be, it is pointless to abstain from buying and eating meat, as the commercial food system is not sensitive to the actions of individual consumers. When I buy a chicken or a hamburger, I do not cause the death of another animal, as my individual spending on meat represents a miniscule portion of the money involved in the industry overall, and it falls far below the threshold of resources required to send an economic signal to agribusiness to kill more animals. In this way, the argument generates the conclusion that eating meat is permissible without challenging any philosophical premises of protectionism.

I have noted elsewhere that the inefficacy objection relies on an overly individualistic understanding of complicity (Lamey 2017). It is a familiar part of moral life to hold someone accountable not only for their individual actions but the actions of groups in which they participate. Julia Driver gives the example of a group of people hiding a body (2016). One of them is physically weak, so that when they all push the corpse into a river, her effort contributes nothing. Even though her individual action makes no causal difference, she can be criticized for participating in the immoral action of the group. Thus even if individual purchasing decisions make no direct causal difference, meat eaters can collectively be criticized in a similar manner. It is not clear, however, that individual purchasing decisions do always fail to make a difference. They can impact the bottom lines of individual restaurants and grocery stores that sell meat (Budolfson 2016a). Given these and other criticisms that can be found in the extensive literature on the inefficacy objection, I leave it to one side.[4]

A second argument I do not address concerns the case for so-called knock-out meat. This proposal would involve genetically engineering food animals so that they do not feel pain (Shriver 2009). This proposal takes seriously the idea that we have a moral obligation to reduce the suffering of food animals and so has something in common with protection theory. But it leaves in place the practice of killing them, which I argue in Chapter 2 constitutes a harm to animals. As even knockout animals would still be harmed by being killed, the proposal is inconsistent with a central premise of protectionism and so represents a form of rejectionism. The fact that in vitro meat offers all the ethical benefits of knockout meat without the drawbacks would also appear to make the development of knockout meat unlikely, providing a second reason to leave it aside.

[4] For the literature on the inefficacy objection as it pertains to eating meat, see the bibliography in Littlejohn (2016).

A third and final question I do not answer concerns invertebrates. The philosophical literature now contains arguments to the effect that a commitment to animal protection entails obligatory entomophagy, the eating of insects (Meyers 2013; Fischer 2016). Such proposals have been defended with arguments that share much with the briefs for new omnivorism that I take up. Doing justice to the question of eating insects would require an additional chapter, as would addressing issues involving milk, eggs, and other foods beyond the flesh of vertebrates and octopuses, the one invertebrate species my account recognizes as sentient. I confine my analysis to arguments for eating meat in the name of animal protection not only to make my discussion more manageable, but because institutionalized entomophagy has been identified as involving many unknown factors, including environmental and waste hazards (EFSA Scientific Committee 2015). With meat we at least know what the consequences of continued consumption are likely to be.

New Omnivorism in Historical Context

Although my primary concern is with the philosophical cogency of arguments for new omnivorism, the timing of their arrival also bears noting. During the same period in which the above-mentioned arguments and technologies have been subjects of attention (or renewed attention in the case of the logic of the larder), the wider culture has been witness to increasing concern with the treatment of animals. In the United States, for example, animal-related initiatives have become a recurring feature of state and local politics across the country, and their overall trend has been toward increasing regulation, and in some cases abolition, of harmful practices involving animals. In 1999, for example, following pressure from People for the Ethical Treatment of Animals, McDonald's began animal welfare audits of its meat suppliers for the first time, a move Wendy's and Burger King, respectively, followed in 2000 and 2001 (Singer 2000: 166–77). Similarly, in 2008 Louisiana became the last state to ban cockfighting, which made it illegal everywhere in the U.S. (Guzy 2008). The same year saw Colorado pass a law phasing out the use of confinement crates for veal calves and gestation crates for breeding pigs, following similar bans on gestation crates in Florida, veal crates in Arizona, and both in Oregon (American Society for the Prevention of Cruelty to Animals 2018).

A similar trend is evident internationally. The European Union began working toward a long-term goal of phasing out battery cages for hens in

1999, while in 2002 Germany amended its constitution to recognize an obligation to protect the interest of animals (European Union 1999; Connolly 2002). In 1999 New Zealand announced a ban on the use of great apes in research, testing or teaching, except when such use is the interest of the ape or its species (and the harms do not outweigh the benefits) (Taylor 2001). In 2006, The Party for the Animals became the first animal-focused political party to win seats in a national legislature when it won two seats in the Dutch parliament, while in 2013 India not only prohibited testing cosmetics on animals but also said that it would not permit the introduction of captive dolphin displays (Álvarez-Rivera 2018; Dhar 2013; Chauhan 2013). Every year, more countries announce similar laws, limiting or abolishing practices that within living memory were considered perfectly acceptable.

From the point of view of what animal protection advocates hope to achieve, changes such as these are only a beginning. Nevertheless, the early years of the twenty-first century may mark an important shift in humanity's relationship with animals. Practices such as meat-eating or wearing leather have endured for so long in part because they have long seemed part of the natural order. This status is well captured by historian James Davidson. A belief, Davidson notes, "can become more rather than less powerful through being unspoken, aspiring to the rank of habit rather than ideology, and a status beyond language, questioning, and argument" (1997: 23). The turn of the twenty-first century was when many practices involving animals began to systematically lose their rank of habit and their aura of unspoken authority, not merely in the eyes of philosophers but in the culture at large.

On the one hand, this historical trend suggests that contemporary society increasingly pays lip service – and sometimes more than lip service – to the idea that animals make serious moral claims on us. On the other hand, it is doubtful there has ever been a time when people were deeply receptive to calls to stop an activity as ingrained as meat-eating, a practice most of us are taught in childhood and go on to engage in on a daily basis as adults. That we live today in a culture of consumption, culinary and otherwise, is unlikely to make our age especially receptive to such a call. The appeal of new omnivorism thus may well resides in its potential to reconcile two otherwise divergent aspects of modern culture. So long as we employ the right philosophy or technology, it promises, a piece of meat can satisfy both our moral impulses and our gustatory ones.

If my arguments are correct, the case for new omnivorism is more often wrong than right. But even so, there is something encouraging about the shift in the animal debate that new omnivorism represents. Carefully formulated briefs for rejectionism, such as the argument of Elizabeth Anderson that I examine in Chapter One, continue to appear (see also Sher 2016: 74–91). But if these arguments are giving way to critiques that take up protectionism's own premises, this may be a sign that such premises are hard to refute. Or so I hope the reader will agree after reading this book.

A Note on Terminology

If the debate over the implications of protection theory have changed, so too has the terminology. In one way this is straightforward. Since Singer and Regan first wrote, traditional protectionists have gone beyond vegetarianism and endorsed veganism (Varner 1994; McPherson 2016; Hooley and Nobis 2016). I follow this shift by referring to veganism rather than vegetarianism, making an exception when I am discussion someone who refers to vegetarianism. As I explain in Chapter 3, so far as veganism is justified by reference to animals (as opposed to other possible concerns such as the environment or health) it is best understood as a diet that is based on a commitment to reducing animal harm. There are ongoing debates as to whether veganism permits consuming honey, eggs from backyard chickens, and other animal products. My concern is not to outline a definitive list of what food items are or are not vegan, important as such a list would be, and so I leave these debates aside to focus on the central claim of new omnivorism, namely, that there is a vegan rationale for eating meat. My goal is to shed light on the ethics of eating animals themselves rather than milk, eggs, and other byproducts.

Given this understanding, I hesitated before settling on the term *new omnivorism*. The arguments I examine could just as easily be characterized as defending new forms of veganism. Indeed, in the case of Davis's argument, the resulting diet is very similar to traditional veganism in terms of the particular food items it involves, hence the term *burger veganism*. By referring to the proposals I examine as a form of omnivorism, I seek to employ a shorthand term that clearly and collectively distinguishes them from traditional veganism. My label draws attention to the central fact that makes them of interest, namely that they either permit or oblige the consumption of meat. Arguments for new omnivorism have received an enthusiastic reception in the popular media, which never fails to emphasize their justifications for meat eating, a feature my terminology highlights (Corliss 2002; Pollan 2002; Bruckner 2015; Smith 2016).

Someone might still challenge my usage of the "omnivore" label on the grounds that it is a scientific term rather than a cultural or political one. Omnivorism, on this understanding, refers to the physiological ability to eat meat, an ability possessed by all human beings, including traditional vegans. For this reason, some protectionists prefer *carnism* as a term to describe the actual practice of eating meat. A worry they may have is that my usage of omnivorism risks giving meat eaters a biological alibi: "Don't blame me for eating meat, it's a biological fact of nature."[5]

For better or for worse I have chosen to use *omnivorism*. One reason is because general usage now deploys it both as a biological and a cultural term. In addition to being more familiar it also strikes me as less polemical than *carnism*. As for the possibility of giving meat-eating an inadvertent alibi, the labels that any work employs need to be read in the context of the work itself. It should be clear that this one challenges rather than defends meat-eating.

A separate terminological challenge is posed by the term *animal rights*, which has come to possess two distinct meanings in the debate over the moral status of animals. It has a narrow usage denoting ethical theories, such as Regan's, that explicitly extend the concept of rights to animals. But it also has a broader use as a political and philosophical umbrella term, to denote any advocacy claim on behalf of animals stronger than that supported by the traditional notion of animal welfare. The broad usage of *animal rights* has value as a familiar piece of shorthand. But this usage is often applied so as to group together all theories of animal ethics that go beyond animal welfare. This is potentially misleading, as many theories of animal "rights" in this wide sense do not actually extend rights to animals, Singer's theory of animal liberation being the most prominent example.

The subtitle of this book uses *animal rights* in the wide sense. Throughout the pages that follow however I generally use *animal protection* as a replacement term. The more familiar term seemed appropriate in the subtitle, which needs to quickly communicate what the book is about to someone coming across it for the first time. But the main text uses the less familiar but more accurate protectionist terminology to avoid exaggerating the importance animal rights play in arguments of many of the writers I discuss, on both sides of the debate over new omnivorism. It has been said that a feminist is a person who believes both that women should occupy a position of equality relative to men, and that society currently denies women such equality. Animal protection is a label similar to feminism in

[5] I am grateful to Josh Milburn for drawing this concern to my attention.

that it combines both a normative claim and a factual claim. A philosophy of animals is protectionist if it avers both that animals warrant a high degree of moral standing and that such standing is routinely violated.

When it comes to my own argument, I endorse the concept of animal rights in Chapter 4, in the context of defending a version of the doctrine of double effect. In principle, however, someone sympathetic to my account could agree to many of the claims I make throughout the book without necessarily endorsing the idea of animal rights. The concept of animal rights is thus a secondary rather than foundational aspect of my overall argument, an aspect my terminology is also meant to reflect.

CHAPTER 1

The Case for Animal Protection

Introduction

In *The Republic*, Socrates begins his inquiry into the nature of justice by depicting life in a just city. After noting the various occupations such a city would need, from farmers and smiths to travelling merchants and sailors, he turns his attention to what residents of the ideal city would eat. "For food they will make meal from their barley and flour from their wheat . . . They will have salt and olives and cheese, and they will have boiled dishes with onions and such vegetables as one gets in the country" (372).

Glaucon is unimpressed. Noting the absence of fish and other Athenian staples, he dismisses Socrates' diet as one fit for pigs. But Socrates is unmoved by Glaucon's disdain. Socrates points out that his concern is to outline the diet of the ideal city, which by its nature is healthy and not "suffering from inflammation." A city swollen with luxuries, Socrates avers, would contain many things his city lacks. "There will be all kinds of hunters," he observes of the non-ideal city, "[and] swineherds too; we had none in our former city – there was no need – but we shall need them along with all the others for this city. And we shall need great quantities of all kinds of cattle if people are to eat them" (373b).

Plato's civic vegetarianism is a little-noted feature of his theory of justice, and has been called "one of the best kept secrets in the history of philosophy" (Dombrowski 1984). Worth noting here is the rationale for Plato's civic diet. Socrates argued that a meatless city would be better able to maintain peace, on the grounds that a vegetarian city would require less land to feed itself than a city of omnivores. The omnivorous city would encroach on the territory of its neighbours, who would covet their land in return. Thus do we find the origin of war, Socrates concludes, "in those passions which are most responsible for all the evils that come upon cities and the men that dwell in them" (373).

A meatless diet has of course long been a central recommendation of animal protection theory. While it may share a recommendation with Plato, however, its rationale is different. The protectionist case for not eating meat stems from a concern not to wrong animals. If we want to fully understand protectionism, therefore, we need to look beyond its traditional recommendations regarding diet and behavior and take note of its underlying principles. This will clarify and distinguish it from other theories that may also overlap on this or that recommendation, and may even offer their own ethical rationale, as Platonic vegetarianism does, yet are not philosophies of protection. Such conceptual clarity is a crucial prerequisite for examining the diverse immanent critics of protectionism who will be the subject of later chapters. Those critics argue that meatless diets and other familiar advocacy claims do not in fact follow – should rather be detached from – protectionist principles. Understanding such arguments is impossible without first getting clear on what protectionism's traditional principles are.

So what are they then? This question can be answered at different levels of generality. Protection theory traditionally takes for granted our least controversial moral views, such as the idea that other human beings can make genuine moral claims on us. It then seeks to enhance them with further principles of its own. In this way, it is like an application for a phone or computer, which achieves functionality by being added to an existing operating system. My primary concern is with the application not the operating system. I seek to outline two principles that have been common to most versions of animal protection: anti-speciesism and moral individualism. While not every argument for protectionism endorses both claims, the great majority do, and so outlining these two principles are the best place to begin.[1] (Insofar as a critique of protection rejects either of these notions, it will be a rejectionist critique rather than an immanent one.)

Both principles achieve their force by presupposing certain moral claims that are not unique to protectionism. The most relevant of these are the bedrock claims that causing human suffering requires justification, and that moral judgment commonly (but not universally) involves extending equal consideration to the interests of individuals affected by our decisions. I therefore begin my discussion by briefly sketching these more general moral principles, which all versions of protection theory draw on. Anyone who has been following the protection literature will note the absence of

[1] Engel (2000) and Scully (2002) offer protectionist arguments that do not appeal to anti-speciesism.

originality in my outline. This is deliberate. The present chapter seeks not to innovate on, but to summarize, the protectionist framework to which immanent critics increasingly lay claim.

Equal Consideration

In the nineteenth century, stories circulated around Europe about Turkish soldiers tossing Bulgarian infants in the air and impaling them on their bayonets (Matthews 1957: 142). These stories may well have been propaganda. Even so, it is not hard to see why they would generate anti-Turkish revulsion. They were based on the widespread and enduring view that the stories describe an unambiguous moral wrong. It seems reasonable to want a moral theory that can accommodate the everyday judgment that it is possible to commit a direct moral wrong against a baby. If that is the case, it is worth asking what our everyday judgment presupposes about morality.

Intuitive revulsion of the Turkish soldiers scenario likely involves more than one moral judgment. But one idea it does seem to presuppose is that the human capacity to experience suffering (a term I use synonymously with pain) is a legitimate subject of moral concern. This does not mean it is always wrong to cause suffering. But when we inquire into the nature of that concern, we seem committed to the view that causing humans to suffer requires justification.

A good sign that an idea can be taken as a premise in moral argument is that is has an air of banality. The notion that causing humans to suffer requires justification meets this condition. It is worth noting however why the preceding paragraph exudes such a dull air. One reason is that it overlaps considerably with common sense. It is possible to be too devoted to common sense, and I do not endorse the mindless view that philosophical *conclusions* must always support prevailing views. But when it comes to *premises*, acceptability is a virtue. The notion that causing humans to suffer is something that requires justification, is one of the least controversial moral claims imaginable. Such a claim loses none of its obviousness when we have babies in particular in mind. They do not need to individually possess something above and beyond the capacity to suffer, for us to require justification for hurting them.

I take it for granted that any minimally plausible theory of morality that applies to our dealings with our fellow human beings will need to take on board the view at hand, namely that causing human suffering requires justification, even when the human has the cognitive abilities of an infant, including someone stalled at such level of ability due to mental handicap.

If that is the case, then a question arises about how causing suffering can be justified. Here again, however, a familiar answer is possible. On our everyday view, we commonly employ a standard of equal consideration when deciding what actions are and are not justified.

Were we forced to choose between saving one human life and saving five, for example, our everyday view says that all else being equal, it would be better to save the five. This is an impartial judgment in that it grants equal weight to each individual life, and judges the action that saves more to be morally superior. The importance of equal consideration also seems evident when considering racism and sexism, a point famously argued by Peter Singer. When a racist grants preference to the interests of a member of his or her race, the wrong he commits would seem to involve a violation of equal consideration. An extreme racist who denies that a particular group warrants any respect at all is indulging in a gross violation of impartiality by refusing to treat the group in question as equals. But even less extreme forms of discrimination are often condemned using a language of equality, one that holds that all social groups warrant the same degree of consideration of their legitimate moral interests.

It is relatively uncontroversial to say that we often make moral judgments regarding our fellow human beings that employ a standard of impartial consideration. Equality in this sense is not a claim that all humans are equal in their abilities, which is obviously not true. Nor is it a claim that everyone should always be treated exactly the same way. It rather has to do with equality in a more abstract sense, at the level of moral consideration. Singer summarizes such a view of equality as the belief that, "the interests of every being affected by an action are to be taken into account and given the same weight as the like interests of any other being" (Singer 1990: 5). When the beings are human, or some sub-set of humans, at least, the use of an impartial standard of consideration is one that often comes naturally.

Singer goes on to note the widespread agreement among moral philosophers that when it comes to our dealings with human beings, the equal consideration of interests is a necessary preconception of any theory of morality, even though philosophers naturally disagree as to what particular measures equality entails (Singer 1990: 6). Singer's remark that a belief in the importance of equal consideration is widely shared echoes an observation many other philosophers have made (Hare 1972: 167, Taylor 1989: 64, cf. Nagel 1979: 111). The implications of this widely held (but of course not universal) belief in equality have been spelled out in the most detail by Ronald Dworkin and Will Kymlicka. Singer is a utilitarian while Dworkin and Kymlicka are deontological liberals, but they uphold the same abstract conception of equality as does Singer. As Dworkin summarizes it, it is the idea that "the interests of the members of the community matter and matter equally" (Dworkin 1983: 24). Dworkin and Kymlicka note how this same abstract conception of equality is endorsed by most contemporary political philosophies. Theories that have been identified as endorsing this abstract conception of equality include not only utilitarianism and liberalism, but feminism, Marxism, libertarianism, and communitarianism, among others (Kymlicka 1990a: 4).

Dworkin and Kymlicka emphasize the role equal consideration plays in the branch of morality having to do with political life and public institutions. Singer, for his part, sometimes seems to suggest that equal consideration is a necessary condition of *any* moral judgment, whether having to do with public life, private life, or anything else (Singer 1990: 5). This is controversial, as it seems to cut against strong intuitions regarding special obligations to our friends and family members. If I were on a cruise with my wife and the ship sank, and I could save either my wife or a stranger, everyday morality says I would be not merely right, but obligated, to save my wife (Williams 1976: 214). This seems to suggest that moral judgments based on certain forms of partiality are legitimate. But if that is the case, it raises a larger question about how or even whether our impartial and partial judgments can be reconciled into a seamlessly consistent whole.

The daunting nature of such a challenge has been well characterized by Thomas Nagel:

> The hardest problems of political theory are conflicts within the individual, and no external solution will be adequate which does not deal with them at their source. The impersonal standpoint in each of us produces, I shall claim, a powerful demand for universal impartiality and equality, while the personal standpoint gives rise to individualistic motives and requirements which present obstacles to the pursuit and realization of such ideals. . . . It emerges that a harmonious combination of an acceptable political ideal and acceptable standards of personal morality is very hard to come by. (1991: 4)

Nagel is highlighting the extraordinary difficulty of fully reconciling impartial and partial moral considerations in a way that respects the manner in which everyday morality draws on both. This truly is an awesome challenge, one Nagel suggests has yet to be met, and one I will not even attempt to meet. While my account stresses the importance of impartial consideration, I do not wish to rule out the legitimacy of partial judgments based on the claims of family or friendship. Protection theory treats a standard of impartial consideration as a necessary element of a theory of morality, but it does not claim that it must be the *only* standard morality ever employs. There may well be a second standard involving partiality that generates some of our moral duties. Alternatively, there may be some way of doing justice to the power of family and friendship ties using an entirely impartial framework (although I doubt it). Regardless, settling the issue of the relationship between impartiality and legitimate partiality would require outlining a comprehensive theory of morality, which is beyond the scope of a basic summary of protectionist principles.

Protectionism only needs to say that impartiality is a standard we do often draw on, and must be included among the elements of any minimally plausible theory of morality, regardless of how a full theory goes on to address the claims of partiality. Generating claims involving impartiality, as protection theory does, will still be knowledge worth having, even if turns out to be knowledge that does not apply to every last moral judgment we make. This is especially so given that impartiality is a standard we do not just employ in political life, but also sometimes in the personal realm (despite Nagel's use of "personal" as a synonym for "partial"). When I am marking student papers, for example, I am a better marker to the degree that I mark all students in the class by the same standard. If I pass a student who should fail on merit because I am friends with her parents, I am indulging a partial tie in an illegitimate way. If I similarly make the racist judgment that I will rent an apartment to a fellow white person but not an indigenous person, even though an indigenous person meets all the criteria in my rental application, I am again granting illegitimate weight to partiality on a personal level. Impartiality in this sense involves not only applying a consistent standard, but one that focuses on relevant rather than irrelevant considerations, ignoring such things as the family background of university students or the race of rental applicants.

Following Singer and Dworkin, I will use the language of the equal consideration of interests to capture this view of impartiality. The fact that both a utilitarian and a deontological liberal use the term "interests" should remind us that the word "interests", by itself, does not entail a commitment to either side of the deontological-utilitarian divide. As David DeGrazia has observed, "seen from the proper perspective, utilitarianism and animal-rights views appear far more alike than different. Crucially, both extend to animals a principle of equal consideration. Any such principle requires that we (in some significant way) give equal moral weight to comparable interests, regardless of who has those interests" (1998: 112; cf. Kymlicka 1990b: 112). It is the underlying commitment to equal consideration, common to deontology, utilitarianism and other theories, that protectionism presupposes. Such a commitment to equality is arguably one that any minimally plausible theory of morality must endorse. It is helpful to conceive of equal consideration in terms of interests because doing so enables us to be precise in our application of the underlying notion of equality.

The equal consideration of interests permits treating one person differently from another when the first person has a legitimate need and the other does not (Singer 1990: 2). Suppose I propose a policy of giving new parents a lump sum payment that is not also given to childless couples. My rationale involves the fact that parents have a legitimate need, attending to the welfare of their children, that is not found among childless couples. Saying that the parents have an interest the childless couples lack is a simple way of expressing this, and is in keeping with the underlying commitment to impartiality (according to which, if one of the childless couples has a baby, they should receive the same payment).

To sum up, I have taken as premises two claims that are uncontroversial when applied to our dealings with our fellow human beings: that a capacity for suffering is a morally salient attribute and, as such, causing it requires justification; and that there is a zone of moral life that employs a standard of impartiality, a standard that is here expressed by reference to the equal consideration of interests.

Anti-speciesism

What does it mean to say that species has no moral significance? The idea can sound like a parody of sexual and racial equality when it is not carefully outlined. When we consider traditional civic and political rights, which are often justified in moral (as opposed to merely legal) terms, they protect the right to vote and other entitlements that would be absurd to extend to animals. Anti-speciesism does not deny this obvious fact. What it does deny is that the brute fact of being *Homo sapiens* matters morally.

There are good grounds for distinguishing our treatment of most human beings from most animals. Most normal adults are persons in the philosophical sense, which refers to entities that possess self-consciousness and its associated attributes. Persons are rational and moral agents with a conception of themselves existing through time. It is reasonable to recognize that a person in this sense will experience a greater loss through death than a merely sentient being. Sentience here refers to a being that is conscious and has the ability to experience pleasure and pain.[2] A person has medium- and long-term plans and other interests extending beyond the immediate present that merely sentient animals do not. In this and other ways the greater cognitive capacities of an ordinary human adult are such that, were we in a lifeboat and could only save such a human or an ordinary adult dog, it would be right to save the human, all else being equal.

This and other familiar judgments link moral standing to cognitive attributes. Anti-speciesism permits such judgments, as it does judgments based on the ties of partiality already noted. What it rules out are judgments based on the idea that our species also matters. It is possible to imagine a human who is mentally disabled and so will never acquire mental abilities more sophisticated than that of a dog. When we compare *that* human being to a canine, his or her level of cognitive complexity does not differ in any significant way, such that both fall below the threshold of personhood. When anti-speciesism is combined with the doctrine of equal consideration, they entail that we should grant the animal the same degree of moral standing we grant the mentally handicapped human.

Pressure to justify speciesism is generated by the fact that we find no inherent moral significance in categories narrower than species. Consider sub-species attributes such as race and sex. We would not want to say that being of the male sex or white race are attributes that generate some extra degree of moral standing (Singer 1990: 2–10). If these categories of identity are devoid of intrinsic moral significance, what grounds are there to say species is different? If we can extend equal consideration to human beings of different races and sexes, it seems consistent with such a stance to maintain that a human and non-human at similar levels of cognitive ability and who are equal in every way save species membership possess an equal level of moral standing.

One reply that is sometimes made to this argument is that race, unlike species, is a social phenomenon. An increasingly widespread view holds that race is the result of a social process that singles out genuine biological attributes, such as skin color and hair type, and invests them with

[2] I here follow many previous animal ethicists going back to Singer, who noted that this usage of "sentient" is "convenient, if not strictly accurate shorthand for the capacity to suffer or experience enjoyment or happiness" (1975: 8–9). A stricter usage would define sentience as consciousness, independent of the ability to experience pleasure and pain. I follow Singer's usage because it is indeed convenient, and because this usage now predominates in the animal ethics literature.

significance. We do not group together everyone who is over six feet tall with blue eyes under a common label, even though height and eye color mark genuine differences. On a social constructivist view, racial groupings have a similar nature, reflecting socially determined categories as much, if not more so, than biological ones. Insofar as species is a genuine biological category, a defender of speciesism might reply that we should not model our thinking about species on our thinking about race, as one is more biologically real than the other.

A problem for this view is that the same concerns about the idea of race have been brought to bear against the idea of species. A well-known issue in the philosophy of biology concerns the longstanding problem of grouping species in a non-arbitrary way. Darwin himself remarked that "it is no wonder that there should be difficulty in defining the difference between a species and a variety; – there being no essential, only an arbitrary difference" (quoted in Ereshefsky 2011: 68). Since Darwin's time much energy has been devoted to addressing the so-called species problem, which stems in part from the fact that scientists working in different areas commonly employ different species concepts, some based on the observable traits of living things and some based on genealogy. The notion that species is a more purely biological category than race is thus controversial. If anything, the grounds to doubt the deep biological naturalness of race also raise doubts about solving the species problem in a way that conforms to our everyday views about the naturalness of species. Moreover, even if there is some way to show that species has a naturalness race lacks, it remains the case that sex (as opposed to gender) is a biological difference, yet not one we view as grounds for morally ranking sexes. In short, even if species is a natural rather than social category, this does not suffice for granting a higher moral standing due to species membership.

Pressure to reject speciesism comes from a second direction. Species is but one category in the hierarchy of biological classification, along with genus, family order, etc. As with categories narrower than species, we do not normally think that genus and other broader categories possess inherent moral significance. To take but one example, human beings, orangutans, gorillas, and chimpanzees are all members of the family Hominidae. Would it be plausible to say that this common membership is morally significant? It is hard to see why this might be. It seems only consistent with our background evaluation of other biological categories to say that species membership also lacks moral significance. Alternatively, if there are grounds for saying that species membership has a moral significance that narrower and wider categories lack, the onus is on the defender of speciesism to outline them.

One conceivable way of making the case is a bedrock appeal to intuition: we just have a pre-theoretical commitment to the moral relevance of species membership that has no equivalent in the other cases. While appeals to such an intuition can be found in the anti-protectionism literature, they are not to the point (Carruthers 1992: 71). To be effective, an appeal to intuition must involve a bedrock intuition shared by the person at whom the argument is directed. Philosophers who reject speciesism, however, do not have the bedrock intuition in question. Invoking it will therefore do no work in the context it is most needed, that of rebutting anti-speciesist arguments, any more than appealing to an intuition about the permissibility of abortion, say, will be germane to a debate about the ethics of abortion. In both cases, the intuition in question concerns the very subject in dispute.

A more theoretical defence of speciesism appeals to evolution. According to proponents of this view such as Neil Levy, there is overwhelming evidence that morality or something like proto-morality evolved "the better to promote the interests of our 'selfish genes'" (Levy 2004: 216). The selfish gene theory is controversial, and Levy takes pains to disassociate his account from some of its more extreme expressions. Morality, he notes, is not just disguised selfishness. "Nevertheless, the rootedness of morality in our evolutionary history might give us reason to extend the highest form of moral consideration to members of our own species, whatever their [individual] properties. It may be that our moral emotions, the building blocks of all moral responses, are inevitably triggered by conspecifics," conspecifics, of course, being members of our own species (ibid., 216).

This defence of speciesism draws a link between morality and the emotions our evolutionary history has hardwired into us. There may well be some sort of link between evolution and morality, but it seems unlikely that appeals to evolution will settle contentious debates between philosophers, showing one side to be clearly right and the other wrong. Both proponents and critics of speciesism are members of the same species with a common evolutionary history, after all, yet they disagree philosophically on a point of moral theory. This suggests that the particular moral view Levy is concerned with is not in fact exhibited by all members of our species (if it were, it is not clear how a debate over the ethical significance of species membership could arise in the first place).

There is a second problem with the evolutionary argument for speciesism. It is not in fact clear that any emotion or reaction can be morally legitimized by reference to its "rootedness . . . in our evolutionary history" (Levy 2004: 216). Anger and sexual jealousy for example are strong emotions that might well be evolutionary adaptations. But if this turned out to be the case, it would hardly follow that judgments made in moments of anger or jealously are morally justified. Even if Levy were correct that our emotions are inevitably triggered in a certain way by conspecifics, therefore, it is not clear that this would count as support for those emotions being morally apt.

It is possible to offer a more sophisticated defence of speciesism than a brute appeal to intuition or evolution. The best defence of speciesism I am aware of is made by Elizabeth Anderson. Anderson asks us to imagine a chimpanzee that can be taught a language up to the level of a young child, as some studies have suggested may be possible. There are handicapped members of our own species who possess a similarly limited capacity for language acquisition. It is uncontroversial to say that handicapped humans possess the moral right to acquire language to the maximum degree their abilities allow. But even though a chimpanzee might possess the identical capacity for language acquisition, up to the level of a toddler, it seems far less wrong to deprive the chimpanzee of language lessons (Anderson 2004: 281; Nussbaum 2007: 364). If that is the case, then according to Anderson, species membership can be relevant to moral judgment after all.

Anderson's response to protection theory is rejectionist. A question her argument raises is, even if it is sound, just how morally significant does it show species to be? It does not, for example, show membership in the human species to be a necessary condition for possessing moral standing. The primitive language scenario does not entail that it would be acceptable to torture a chimp for our own amusement. Thus even if species is a morally salient characteristic, it does not follow that being a member of our species is either necessary or sufficient to possess moral standing.

Similarly, our judgment in the chimp language scenario might change if we altered some of the details but kept the difference in species. What if we were comparing a normal human infant and a remarkable chimp, both of which had the potential to learn full competency in a language such as English? Would we still make two different judgments, or would we say that both now deserve language instruction? It seems consistent with the chimpanzee language argument to say that in this case we would in fact commit an equal wrong by denying either entity full linguistic ability. Alternatively, there could be less important capabilities that it would be acceptable to leave unactualized in the case of both chimp and human. Suppose I have the capacity to be the world's greatest yacht racer, but lack the resources necessary to actualize my potential. Does society do me wrong by not affording me yacht racing lessons? The intuitive answer is no. If a super-chimp someday appeared that is capable of captaining a yacht, surely we would do no wrong in not training it for the America's Cup either. If so, our judgments about species as it applies to partial language acquisition may not be generalizable beyond the particular scenario that Anderson mentions.

But there is a deeper issue. In the course of making her argument, Anderson distinguishes "Species membership per se" from "species life," which refers to the behaviors and forms of life that it is normal for members of a particular species to engage in (Anderson 2004: 283, 281). Human beings, for example, typically live in society with other humans, a society to which wild animals do not belong. Anderson is at pains to show that there are moral entitlements that extend to members of society that can be legitimately denied wild animals, on the grounds that they are not members of society. This claim is reasonable, but it is conceptually distinct from a claim about species membership per se, which her chimpanzee language example is meant to show. And once the distinction between species membership and species life is made, it should prompt us to ask whether the Chimpanzee language argument does in fact show that species *membership* is morally important even in the limited domain it treats.

We can answer this question by imagining a chimp and human in reversed situations of species life. There was once allegedly a chimpanzee named Oliver, for example, who lived with a family of animal trainers and exhibited many human-like behaviors. "Oliver occasionally fed the dogs and did other chores, relaxing afterward with a cup of coffee," one magazine account reads. "In the evening he might sit and watch TV with the couple, sometimes preparing a nightcap for Frank and himself of whiskey and 7UP" (Shreeve 2003: 1). If we imagine an Oliver-style chimp, one that has already been deeply integrated into a living arrangement with humans, and who would truly benefit from learning language to the same degree as would a mentally handicapped human, then the case for teaching primitive language skills to the chimp seems as strong as in the case of the handicapped human. For the chimp's life is now one strongly defined by relationships with humans. Insofar as those relationships will be strengthened and deepened by obtaining limited language skills, and the chimp would be better able to communicate his needs and participate in a wider range of behaviors and activities, it is not clear why he should not be entitled to such goods to the same degree as a human with identical cognitive capacities in the same situation.

Conversely, we have all heard stories of feral humans, raised by wolves or some other animals, whose behaviors they mimic. So let us imagine a world in which there has been some catastrophe wiping out the entire population except for one child. Let us suppose that the child is mentally handicapped to the point that her cognitive abilities can never develop beyond the level of a chimpanzee. And now let us suppose that the child is adopted by chimps and is raised to live among them in the wild. The grown child will no more benefit from coming across a device that will allow her to obtain limited fluency in the language of her former civilization any more than a fellow member of her chimpanzee troop would in stumbling upon the same technology. In both cases, whatever benefits there are to limited human language acquisition will be experienced equally by the isolated feral human and her fellow chimp. If that is the case, Anderson's example does not in fact demonstrate that species membership per se, rather than species life, is morally salient. Indeed, when the integrated chimp and the feral human in the science fiction scenario are compared, I would argue that the integrated chimp, if anything, has a stronger case for limited language instruction, even though she is not human. If that is the case, then the chimpanzee language argument does not succeed in demonstrating the moral relevance of species membership.

It is no threat to protectionism to say that beings with different species lives may have different interests. The goal of protection theory is not to show that animals share every last human interest, but to show that we should grant equal consideration to those interests that do cut across species, such as an interest in avoiding suffering. It is consistent with protectionism therefore to allow that a chimp living in the wild and a human being at a similar level of cognitive development living in society can have non-identical interests. This however is not because of their different species, but because of their different circumstances. And it is doubtful that this difference is captured by the notion of species-life.

The idea that there is a human community with its own species life, distinct and separate from that of the species lives of all animals, is not quite right. The lives of human beings intersect with those of animals on a daily basis. Interacting with a dog or other domestic animal, for example, is part of a recognizably human form of life, just as interacting with us is part of a domesticated animal's form of life. In addition, there are many species of animals that fall somewhere in between the wild-domestic distinction. The forms of life of many raccoons, squirrels and pigeons, for example, involve interacting with human beings in an urban environment.

Will Kymlicka and Sue Donaldson have suggested that we take seriously the thought that many human and animal lives will go best in communities and forms of life based on interaction across species. "Our conception of human flourishing shouldn't assume that our most significant relationships must be with other humans, as opposed to other species," Donaldson and Kymlicka write. "To pathologize these preferences as deviating from an alleged human norm is to close ourselves off from the potential richness of interspecies sociality" (2011: 99). What is true of human beings is true of many animals as well. "For a dog or donkey at a rural farm sanctuary, relevant capacities might relate to getting along with a variety of other animals (of varied species), education about the hazards of farm machinery, or the learning of useful skills like protecting sheep or shooing crows from the corn crib" (ibid., 97). Still other forms of flourishing would apply to a domestic dog or cat. In short it is possible to exaggerate the degree to which forms of life are determined by species membership. Some forms of life, including the ones many humans and animals currently practice, are made possible by living and interacting with more than one species. The notion of "species life" is misleading to the degree that it obscures the everyday reality of forms of life that involve interactions between species.

Species life is additionally misleading when it obscures the reality of forms of life that are shared by different species. The existence of such forms of life is suggested by Dawn Prince-Hughes, a person with autism who developed a deep identification with gorillas after observing them for hours at the Seattle zoo. Prince-Hughes's experience is similar to that of well-known autism spokesperson Temple Grandin, whose method of slaughter we will examine in Chapter 6. Both writers describe a sense of comfort in being around animals, which contrasts sharply with the high anxiety they commonly feel in social situations with other human beings (Grandin 2006a). Prince-Hughes was able to relieve her anxiety after getting her a job as a zoo helper, which allowed her to enter the gorillas' enclosure and closely interact with them. Through careful observation, she learned how to negotiate social life as a person with autism:

> I tried to apply things I'd learned from the gorillas in social situations. I tried to put people at ease by acknowledging them with quick sideways glances and smiles – which evolved from submissive primate grimaces and are intended to convey that no harm is meant. I learned how to avoid arguments by putting my hands up with the palms facing outward and head down. I asserted myself by making myself bigger, putting my hands on my hips and pushing my elbows forward .... By applying the bodily and verbal language

> components I had learned from the gorillas, I was beginning to have more social success. (Prince-Hughes 2004: 136–7)

Prince-Hughes was able to overcome challenges to her flourishing by taking cues from members of another species. What she learned was not a minor new skill, but a basic ability to socialize with others. This suggests that Anderson's notion of species-life exaggerates the degree to which forms of life are species-specific.

The notion of species-life will not be saved by switching our attention away from Prince-Hughes and looking for examples of species life that are particular to people who are neurotypical. On such an approach a candidate for species-life status would be something like the ability to pass the Salle-Anne test, used by developmental psychologists to test a child's ability to attribute false beliefs to another individual, and which autistic children pass at a far lower rate than other children (although there is debate as to why: it was initially attributed to cognitive inability, but may in fact be due to social anxiety). By defining species-life in this way, we would simply be admitting that it does not track species membership at all closely. Any notion of species life exclusive enough to exclude Prince-Hughes will exclude millions of other people with a wide variety of handicaps and disorders, yet who nevertheless remain *Homo sapiens*. Our concept of species life would now be so restrictive that only confusion could continue to justify retaining "species" as part of its name.

Shifting to a normative conception of species life – one that holds up certain forms of life as those that *should* be distinctive of human beings, even if they are not – will not work either. Insofar as such a theory treats human beings who are unable to follow a preferred form of life the same way it treats animals who also cannot do so, it will pose no challenge to anti-speciesism. Insofar as such a theory invokes species membership as a basis for treating the two cases differently, it must justify why such a distinction is relevant. Without such justification, this merely begs the question, and brings the discussion back to square one.

Moral Individualism

With speciesism ruled out, what does it mean to extend equal consideration to sentient animals? Such animals are now correctly understood as occupying a position of moral standing equivalent to that of human beings at an equivalent level of cognitive ability. It is no more acceptable to subject animals to practices such as factory farming, for example, than it would be to use certain mentally handicapped individuals for the same purpose. The interest of the animal in not being subject to such practices is strong, while the interest gained by hurting them is trivial: there is no great moral urgency requiring us to eat factory farmed meat when other alternatives are widely available. Only by granting our own pleasure and convenience excess weight could we justify causing animals to suffer for such purposes.

A view based on equal consideration and anti-speciesism by itself does not answer every moral question involving animals. The ethics of killing, for example, is a subject we will take up in the next chapter. But it does suffice to immediately call into question a wide range of practices involving animals currently deemed acceptable. Unless we want to achieve consistency by subjecting some mentally handicapped humans to the same painful practices we currently inflict on animals, then animals who occupy an equivalent degree of moral standing require much better treatment than they currently receive.

Questioning the discrepancy between society's treatment of human-non persons and its treatment of animal non-persons has long been an established protectionist strategy. Insofar as the targets of protectionist arguments have any sympathy for human beings who fall below the threshold of personhood, the comparison to animals will highlight an unjustified double standard. A possible response to the comparison however is to deny any such equivalence, on the grounds that the mentally handicapped are part of a larger group, human beings, who *typically* have a trait that generates high moral standing, and which animals lack. Examples of such a trait include personhood, moral agency, a sense of justice or rationality. Whichever trait might be put forward, the high standing generated by possessing it can potentially be extended, critics of protectionism have argued, to the mentally handicapped (Cohen 1986). This is by virtue of their membership in the species *Homo sapiens*, of which possessing the trait in question is typical.

While different theories may nominate different traits as the grounds of higher moral status, the overall structure of the argument that seeks to work from them to include the severely mentally handicapped is the same in each case. This underlying structural affinity has been correctly identified by Neil Levy, who puts forward the principle of derived moral significance to capture the idea common to all such attempts at extension:

> If (1) an individual A is a member of some species S and (2) some, most or all of the other members of that species have some property C and (3), on the basis of having property C, they have moral property R, then individual

A has moral property R as well, even though A lacks property C (Levy 2004: 215).

Suppose individual A is a member of the human species who is so congenitally handicapped that they lack the traits of personhood. Suppose also Property C is personhood, widely regarded as a privileged trait that bestows advanced moral standing, which would itself be moral property R. If the principle of derived moral significance is correct, the path is cleared to bestowing a kind of honorary moral status on the mentally handicapped individual, even though they lack the property that normally grounds higher status. They now can possess the moral status of persons, which continues to be denied to animals.

Levy's principle accurately characterizes a move that was once common in the animal ethics debate (White 1989; Cohen 1986; Schmidtz 1998; cf. Norcross 2004). The appeal of such a move is not hard to see. It denies the claim of equivalence protectionists have argued for regarding the moral standing of humans and non-humans who differ only in their species membership. The problems with such a move, however, are now well known. No matter what trait we identify as bestowing high standing (property C), all such moves will run afoul of moral individualism, which any minimally plausible theory of morality must unambiguously affirm (Rachels 1990: 5). Indeed, the truth of moral individualism represents an insurmountable problem for any approach that attempts to reject it.

Because of its nominal similarity to political individualism, moral individualism is easily misunderstood. Political individualism stresses the importance of the individual at the expense of the group. Traditionally, it has been invoked to challenge the importance of collective traits such as gender, class, or race. Moral individualism is very different. It does not have any partisan political implications, and is compatible with arguments for and against doctrines such as feminism, multiculturalism, and socialism. Moral individualism makes the claim that an individual entity's attributes should determine its moral entitlements. It does not take a stand on which attributes, collective or individual, are in fact important. It merely holds that moral judgments that stress any given trait are only applicable to individuals who do in fact possess that trait.

To illustrate, imagine a society in which English speakers are disadvantaged in some ways and advantaged in others. On the one hand, English speakers are in the minority and their language is under threat. The situation is so bad that the government has established a separate

English-language school system in order to help preserve the language. On the other hand, English speakers have higher incomes on average than the general population. Under the society's progressive taxation scheme, Anglophones pay more taxes on average than other people do.

Now suppose I am an English-language parent who makes substantially less money than other Anglophones. What would moral individualism rule out? It would not rule out sending my children to an English school. My status as an English speaker is unambiguous, and insofar as society has determined that this is the criterion for enrolling my children, I qualify. To be an English speaker is to be a member of a group, so it should be clear that moral individualism takes no stand as to whether making judgments based on group membership is acceptable. (To argue for or against a separate school system for Anglophones, one would have to invoke considerations other than moral individualism.)

What moral individualism would rule out is taxing me at the same level as most Anglophones. While I belong to the Anglophone group, I do not share the particular trait on which the higher tax rate is based, namely, having a higher income. The judgment involved in the school system was a simple and direct one. Speak English, enroll your children. In the tax case however the reasoning follows a more indirect route. To the degree that one's income is higher, one pays a greater proportion of taxes. Anglophones *as a group* have a high average income. I am an Anglophone. Therefore, I should pay the high tax rate. Moral individualism rules out this second method of grounding moral judgments. Insofar as any property is taken to ground a moral entitlement, that entitlement can only be possessed by individuals who do in fact exhibit that trait. Whether one shares some other quality with people who have the trait in question is irrelevant, just as it is irrelevant that I speak the same language as the group that pays high taxes.

The application of moral individualism to human beings is uncontroversial as a moral principle. It is compatible with a wide range of moral and political philosophies (including those that would oppose separate school boards for linguistic minorities). Indeed, at the level of our everyday morality, it is quite common for us to base our judgment about an individual on those traits, unique or shared, the individual actually possesses. The idea that it would be wrong to send an innocent person to prison, for example, or wrong to make you pay my debts, are only some of the many judgments we frequently make that are informed by moral individualism. All moral individualism rules out is basing judgments about an entity on attributes that the entity in question does not possess.

This however is enough to make it significant in the present context. For it does rule out attempts to bestow an honorary moral status on the mentally handicapped individual generated by their membership in a group whose *typical* members have greater cognitive abilities. Moral individualism instead insists that the standing of the disabled individual, like the standing of any moral subject, be derived from their mental capabilities and other attributes they do in fact possess.

One possibility is for defenders of the principle of derived moral significance to stand their ground, and maintain that species membership is unique, the one case in which moral individualism does not hold sway. But this again would require making the case for speciesism, and would again bring us back to square one.

Conclusion

If what has been said up to now is correct, it has significant implications for practices involving animals. Consider three brief examples. The first is hunting for sport. It is perfectly legal in our society to go out into the woods and shoot animals for pleasure. This type of hunting should be distinguished from what is sometimes referred to as therapeutic hunting, or hunting that involves killing some wild animals for the sake of their or another species. When killing some animals is the only way to return an ecosystem to normal functioning, and non-lethal alternatives are not available, an equal consideration account can justify such hunting, as it is carried out as a means of preventing even more animals from suffering (as would occur, for example, if an ecosystem were overrun with members of an imported species that hunted a local species to extinction, disrupting the entire ecosystem).

Hunting for sport, by contrast, is done with an eye for the pleasure of the hunter rather than the interests of the animals. It often involves killing animals that are in fact well adapted to their ecosystem. Hunters commonly take pride in killing the biggest animal possible and leaving the smaller ones, a pattern of intervention that can have negative results for the ecosystem, in that those animals that are most well adapted to it are more likely to be killed. On our everyday morality, we would never say that if someone took pleasure in shooting mentally handicapped orphans, such pleasure would justify doing so. On an equal consideration account, the pleasure that hunters take in hunting is also insufficient justification for the suffering they cause animals when killing them with firearms, crossbows, or similar painful means. On our everyday view, the mere fact that something is pleasurable is not grounds to cause human suffering. On a protectionist account, a similar standard should govern our dealings with animals.

A second practice that is called into question by the equal consideration of interests is animal testing done in order to bring new cosmetics to market. As one account of such testing notes, the pain for the animals can be considerable:

> Product testing is commonly performed on animals to measure the levels of skin irritancy, eye tissue damage, and toxicity caused by various substances used in the manufacture of cosmetics. In the Draize test, caustic substances are placed in the eyes of conscious rabbits to evaluate damage to sensitive eye tissues. This is extremely painful for the rabbits, who often scream when the substances are applied and sometimes break their necks or backs trying to escape the restraints. (IDAUSA 2010)

In such cases, a legitimate moral interest of the animals, that of not experiencing severe pain, is overridden in the name of what seems a fairly trivial interest, namely having a new cosmetic on the market. Certainly it is difficult to imagine any member of our own species being put through such procedures in order to achieve the same goal. Such a reluctance would seem to apply even in the case of humans who are sentient without being self-conscious, as rabbits might be thought to be. If blinding them were the only way to bring new cosmetics to market, it seems likely we would say that we already have enough cosmetics, and having more to choose from is not worth the cost in suffering. If that is the case, an equal consideration account obliges us to take a similar view of cosmetic testing on animals.

A third and final practice called into question by the equal consideration of animal suffering involves intensive animal agriculture. By intensive animal agriculture I mean farming systems that cause animals to suffer significant pain. Animal suffering has traditionally been widespread throughout industrialized farming practices. When chickens are tightly housed together in large numbers, for example, they begin to peck at and even eat one another. The traditional solution has been to "debeak" them, or remove their beak without the use of anaesthetic. A similar lack of concern is evident in the manufacture of veal, which sees calves isolated from their mothers and housed in crates so small they cannot turn around, where they are denied iron and other nutrients, in order that their flesh might take on a preferred tenderness. The result is that they often succumb to disease, stress and cuts from rubbing against the stall. Kosher and halal slaughterhouses similarly require that food animals be killed with a single slice of the knife, and not be stunned prior to killing, which enhances the pain the animal experiences (Grandin and Regenstein 1994). These painful practices are also unethical on a protectionist account. This, however, still leaves unaddressed the possibility of killing animals painlessly. Addressing that possibility requires investigating the ethics of killing distinct from suffering, which is the subject of the next chapter.

CHAPTER 2

A View to a Kill

Introduction

Moral claims regarding animals inevitably presuppose factual claims about their mental abilities. The notion that we have an obligation to minimize animal suffering, for example, takes for granted that there are members of at least some non-human species that feel pain. In the Western philosophical tradition, a recognition that animals can suffer stretches back to ancient Greece and is today reinforced by evolutionary theory, which classifies the capacity to feel pain as an adaptation common, at a minimum, to all vertebrate species.[1] The claim that suffering can be found beyond the membership of our own species is now less controversial than its denial, and its wide acceptance is one reason why the previously noted trend of rising moral concern for animal suffering could occur.

But what about killing? Does causing an animal to die, as distinct from causing it pain, harm it? A widespread view holds that it does not, due to the limited cognitive abilities of animals. Such a view is articulated by J. David Velleman. He holds that in order for killing to harm an entity, that entity must be capable of caring about what death deprives it of, namely, its own future. Given that animals appear unable to do this, Velleman concludes that killing them is not intrinsically wrong:

> I assume that a cow cannot conceive of itself as a persisting individual and consequently cannot conceive of itself as enjoying different benefits at different moments during its life. What the cow cannot conceive, it cannot care about ... This conception of a lower animal's interests is supported, I think, by its fruitfulness in accounting for our intuitions about the moral

[1] See Sorabji (1995) for debates about animals in the ancient world. See Varner (1998: 53–5) for an informed account of what species feel pain. In making my account I generally have vertebrates, particularly food animals in mind, buy for ease of expression I refer to *animal protection* rather than the more pedantic *vertebrate protection*. But see also the discussion of octopuses in Chapter Four.

> difference between killing animals and killing people. . . . if there is no moment at which a cow is harmed by death, then it cannot be harmed by death at all (Velleman 2000: 81–4).

Velleman's view is thus that animals are not harmed by death per se, a view that has been embraced by many other philosophers (cf. Hare 1999: 238; Thomson 1990: 292–3).

On an empirical level, we might wonder if the mental abilities of any given species are as limited as Vellman's view assumes. In Chapter 6 I will analyze recent findings regarding avian cognition to argue that at least some non-human species have greater cognitive abilities than has traditionally been recognized. My present concern, however, is different. It is to present an alternative view of the ethics of killing animals that takes for granted that their mental abilities *are* as minimal as Velleman portrays them to be. Animals, on the view I am about to defend, can be harmed by death even if they cannot form desires about the future and cannot have a conception of themselves through time. If such a claim is correct, it means that even killing an animal painlessly will require justification, and the purpose for which the killing is carried out will be ethically relevant.

My present goal is to outline the conceptual resources that protection theory brings to bear on the issue at hand, which is killing. These resources will inform the critical engagement with immanent critics of animal protection that begins in Chapter 3, where I will go beyond what other protectionists have already said. My immediate aim, however, is not to present a new view of the ethics of taking non-human life, but to outline the most cogent protectionist view of the subject currently available. Such a view employs the notion of a time-relative interest, which refers to the interest an entity has in continuing to live. This interest is shaped by two factors. They are the gain or loss of future well-being, and the amount of psychological continuity between the entity now and in the future (when its well-being improves or declines). After outlining the time-relative interest account, I note how it differs from another influential ethical theory that takes future well-being into account. This is Don Marquis' famous notion of a future like ours, which Marquis employs to argue against abortion, an implication my account does not share. I then defend the time-relative interest account from the criticism that it fails to grant weight to interests a deceased individual would have come to possess had he or she not died. Such a view is a result of mistakenly generalizing from what it says about killing a fetus to all cases of killing. When the theory is correctly understood, it clearly distinguishes fetal killing from other cases. Finally I discuss the implication of the time-relative interest account for infanticide, and show why endorsing it is consistent with a legal ban on infanticide.

Psychological Unity

The argument that animals are harmed by death employs the concept of a time-relative interest (TRI), a notion developed by Jeff McMahan. This idea is itself part of a larger account of the ethics of killing that presupposes the notion of psychological unity, which refers to the degree to which a self remains the same self through time. In order to understand the time-relative interest account of killing, therefore, it is necessary to first outline the subtle and complex notion of psychological unity on which it relies.

Psychological unity is a concept with more than one component. One of those components is psychological *connection*, which itself can take more than one form. A basic form of psychological connection occurs when a belief or psychological disposition exists at one time and continues to exist at a later time. A second form of psychological connection involves mental states that occur at different times but contain some internal reference to one another. Examples would be an experience and a later memory of it, or forming a desire or intention at one time, and the subsequent experience of fulfilling or acting on that desire.

When it comes to the first kind of psychological connectedness, a mental state simply enduring through time, it would be present to a comparatively insignificant degree in the life of a dog, insofar as the dog retains its various beliefs, dispositions and so on, as when it is hungry or sated one minute and remains so the next. In that sense there is a minimal degree of psychological unity in the dog's life, as type-one psychological connectedness is a component of psychological unity. But two major dimensions of psychological unity are missing in the dog's case.

For one thing, the sheer number of mental states possessed by the dog is minimal. It has far fewer beliefs than a person, for example, and so a person's overall psychological unity will be much, much stronger, due to the sheer density or magnitude of her total beliefs. Additionally, in the case of the dog there is little internal reference among the earlier and later mental states: few memories of discrete experiences, and few actions that satisfy desires or intentions formed at a significantly earlier point in time. Both of these absences sharply limit the degree of overall psychological unity it is possible for the dog to achieve, relative to a person.

When the two elements the dog lacks are added to the first, we see that psychological unity is a notion involving three components. In McMahan's helpful summary, the degree of psychological unity within a life between one time and another is a result of "the proportion of the mental life that is sustained over that period, the richness or density of that mental life, and the degree of internal reference among the various earlier and later mental states" (2002: 75).

An upshot of this account is that a life is something over and above the experiences that it contains. McMahan gives the example of a sentient creature that has even less psychological unity than a dog. The mental life of this fictional creature takes the form of a stream of consciousness without *any* psychological connections. It cannot remember what happened to it one second previously, it is not self-conscious, and it has no sense of the future. It has no beliefs, desires, or dispositions to carry forward through time. All it has are experiences of a certain degree of pleasantness (McMahan asks us to imagine it has a pleasure center in the brain, but little else in the way of neurological development).

Let us call such an entity a sentient oyster. It would have a long stream of pleasant experiences extended before it into the future. But the creature is not related to itself in the future the way a person is related to herself in the future. There are no grounds to care, for the sentient oyster's sake, whether it ceases to exist and is replaced by another oyster with the same capacity for pleasure. For the first creature's stream of internally unrelated pleasant experiences have no continuity between them, and a similarly unrelated stream of experiences will also exist when it is replaced by an identical oyster.

In the case of a person, by contrast, her psychological unity provides grounds *not* to be indifferent to her being killed and replaced by another person. For when the first person dies, a stream of experiences that are internally related cease to exists, and experiences with such a relation cannot be replaced by bringing another person's experiences into existence. A person's future in other words matters for her in a way that the future of the sentient entity "marooned wholly in the present" does not matter for it (ibid., 76).

When a person ceases to exist, she therefore suffers a loss that the sentient oyster does not, and it is for this reason fair to say that a greater wrong is done when the person is killed. For while we deprive both the sentient oyster and the person of a set of future experiences, only the person loses a set of experiences with continuity between them. As McMahan puts it, the strong psychological unity found in the lives of persons, "gives our lives as wholes a moral and prudential significance that the mere sum of our experiences lacks – or, to put it differently, that makes our lives as wholes significant units for moral and prudential evaluation" (ibid., 76).

The notion of psychological unity is conceptually distinct from the concept of future experiences. This is not the same as saying our future experiences are unimportant to us. Of course they are. But there is something in addition to the experiences themselves that gives us a reason to care about our future. It is that they are *our* experiences. This is the difference that psychological connectedness makes. It can perhaps most clearly be seen by considering one of McMahan's thought experiments in which future positive experiences occur but psychological connectedness does not:

> *The Cure.* Imagine that you are twenty years old and are diagnosed with a disease that, if untreated, invariably causes death (though not pain or disability) within five years. There is a treatment that reliably cures the disease but also, as a side effect, causes total retrograde amnesia and radical personality change. Long-term studies of others who have had the treatment show that they almost always go on to have long and happy lives, though these lives are informed by desires and values that differ profoundly from those that the person had prior to treatment. You can therefore reasonably expect that, if you take the treatment, you will live for roughly sixty more years, though the life you will have will be utterly discontinuous with your life as it has been. You will remember nothing of your past and your character and values will be radically altered. Suppose, however, that this can be reliably predicted: that the future you would have between the ages of twenty and eighty if you were to take the treatment would, by itself, be better, as a whole, than your entire life will be if you do not take the treatment. (McMahan 2002: 77)

This scenario is meant to pump our intuition about what matters when assessing our future. Is it the future experiences themselves, or the fact that our current self is the one that is having them? Anyone who would be inclined to reject the cure would be making a judgment consistent with McMahan's emphasis on psychological unity. For in rejecting it one would be rejecting a future in which one was effectively a stranger to oneself, long-lasting though that future is, in favor of a shorter one in which one's present and future selves have a recognizable connectedness.

Now some people may hesitate in deciding between the options in the Cure scenario, and not feel completely sure that they would in fact opt for the shorter yet continuous life. But even hesitating in the face of the options the cure scenario presents can be revealing. For even hesitation raises the possibility that something is lost in choosing the longer, but less continuous, life. To see what is lost, we can reconceive the Cure scenario so that now both lives extend to eighty. The first one in a continuous, normal way, the second one with the same radical discontinuity as already discussed. If we judge these two cases differently, so that continuing to exist with our present values and memories is more appealing than living without them, then we are again granting significance to psychological unity, which is all the Cure scenario is here meant to make us see.

Of course there is a third possible response to The Cure. In response to the scenario as McMahan originally presented it, one might confidently proclaim that the longer discontinuous life is clearly superior than living the shorter continuous one. One possible reason to take this view is that one ascribes positive weight to future good of the kind offered by the discontinuous life, and no weight to psychological unity. The Cure scenario hardly represents a choice at all on this approach, as one path represents something of clear value while the other does not.

The longer discontinuous life will not be attractive to anyone who feels the shorter life contains something of value that the discontinuous longer existence lacks. Certainly in my own case, I do not find the longer discontinuous life an obviously correct choice in The Cure. For this reason, I am inclined to agree with McMahan that psychological unity through time has moral and prudential significance. Our intuitions regarding The Cure, however, are not the only reason to recognize the importance of psychological unity. For there are entities in the real world that possess very weak psychological unity with their future selves.

Such entities include zygotes, fetuses, and infants. As McMahan notes, they do not have a sense of the future or ability to remember into the past, and so cannot form mental states with internal reference to mental states at other times, such as memories. In the case of a zygote, the degree of psychological connection between it and its future is even less than that of a dog – even less than that of the sentient oyster. For it is not even a conscious entity able to experience mental states of any kind, let alone mental states with any degree of continuity through time. But if an entity's future experiences is all that seriously matters, and psychological unity matters not at all, it seems not merely wrong to kill a zygote, but even more wrong to kill it than, say, a 30-year-old person, all else being equal.[2] For the

[2] All else being equal means we grant that the judgment may be different when we have certain knowledge that someone will be tortured or imprisoned for the first 30 years of life, or for some other reason experiences a life not worth living.

amount of future experiences extending before a zygote is potentially significantly longer, given that there are 30 more years in which those experiences might occur. McMahan suggests that the counterintuitive outcome regarding zygotes and early stage fetuses is good grounds to reject the view that all that matters morally in assessing our future is the amount of future positive experiences or future good they contain, with no regard to psychological unity.

In addition to the intuitions generated by The Cure, and the implausibility of ranking the death of a zygote worse than that of a person, there is a third reason to reject the future-good-is-everything stance. This third reason involves the issue of non-conception. As has already been noted, on our everyday morality, we do not consider it immoral to bring into existence fewer people than we are capable of. This is why we do not condemn couples capable of raising ten children when they only have one or even zero children. But if future good is all that matters, to the point that even a zygote is wronged when its life is terminated, it is not clear why non-conception, which also involves a loss of a future good, is not also wrong. Such a conclusion, however, seems absurd. The present account, which emphasizes not just future good but psychological continuity, does not entail such a counterintuitive conclusion. For there is no entity that is wronged by non-conception, let alone one that exhibits psychological continuity with a future self, that is denied its future experience or good.

Time-Relative Interest

McMahan's presentation of psychological unity has subtle details and connections to his overall theory of the ethics of killing that I must skip over here.[3] I have outlined McMahan's notion of psychological unity and three reasons to accept it. But the central relevance of psychological unity for my account concerns the role it plays in the notion of a time-relative interest. As with most theories of morality, the TRI account holds that normal adult human beings have an interest in their future. "Interest" in this sense, of course, refers not to being curious about one's future, but in the sense of having a stake in it. Future events can have an impact on my

[3] One detail may bear pointing out. While McMahan's book does present a theory of personal identity, psychological unity is not a necessary condition of personal identity. There are other grounds, involving the continued physical existence of the brain, that inform McMahan's account of personal identity per se (McMahan 2002: 79). There are hypothetical scenarios in which someone can be conceived as of as having the same identity without psychological unity. The long but discontinuous life option in The Cure is one example.

life as a whole and my overall well-being. That impact on well-being is influenced by two factors.

The first and simpler consideration is the value, negative or positive, the future event will have for me when it happens. If I face a firing squad tomorrow, my overall welfare will be strongly impacted in a negative way, and I can reasonably be said to have a strong interest in my execution not happening.

Not all future events are created equally, however. We also need to give some weight to consideration number two, which is the "prudential unity relation" that holds between oneself now and in the future (McMahan 2002: 81). The prudential unity relation is the degree to which we have legitimate grounds to be concerned, or someone else has grounds to be concerned on our behalf, with the events of our future. This prudential unity relation between a self at one time and another can vary, crucially, according to the degree of psychological unity between the present and future self.

In the modified Cure scenario, for example, whichever path I choose will see a 60-year period that involves and impacts on my interests. But the future in which my beliefs, values, etc. are continuous with my present ones is the future I have a stronger prudential unity relation to, because my psychological unity with my future self is stronger. My well-being is more strongly influenced by this future because it is more truly *mine*.

The concept of a time-relative interest is meant to express this sense of interest in one's future. That interest is one that varies not only according to the sheer scope of future events, my death being worse for me than being sick for a few days, but also according to the degree of psychological unity, and so the prudential unity relation, between me now and me when a future event occurs. Those interests are time-relative in the sense that they take into account my state at a given time. I have a weaker time-relative interest in a pleasant trip to the park that will occur after the onset of Alzheimer's and the substantial disintegration of my psychological unity than I do in a similar trip that takes place before Alzheimer's sets in. Similarly, a week-old infant has a time-relative interest in its future, but it is a weaker interest than the one I have in my own future, as there is less psychological unity between the infant and the adult it will later become than there is between my adult self now and in the future.

To sum up the notion of a time-relative interest, it is the interest I have in a future event, relative to the impact the event will have on my well-being and relative to the amount of psychological unity between my self now and when the event in question occurs. I have already mentioned some of the reasons this concept, and the underlying notion of psychological unity on which it is based, is appealing. One is that taken together, they can explain why opting for the longer discontinuous life in The Cure is not automatically superior. Another is that they are consistent with the notion that killing a zygote or early-term fetus is not wrong, let alone as wrong as killing a person. Third, time-relative interest and psychological unity do not entail any counterintuitive outcomes regarding non-conception. McMahan, however, notes an additional reason to embrace the time-relative interest account that I have not mentioned. It is that the time-relative interest account can capture the idea that, all else being equal, a greater deprivation occurs when someone dies at 30 rather than 80. The loss of future life is greater, and so the scope of the deprivation is larger. The theory's emphasis on psychological unity, however, allows it to incorporate this judgment without entailing the counterintuitive claim that dying at one week old is even worse than dying at 30. For all these reasons, we should recognize the prima facie plausibility of the concept of a time-relative interest.

Time-Relative Interest Versus a Future Like Ours

This concludes my general outline of the time-relative interest account of killing. Clearly, it is a theory that grants weight to an individual's future goods. There are other prominent theories that also have this feature, with which the TRI account could potentially be confused. Before moving on to the specific issue of killing animals, therefore, it may be helpful to note how the time-relative interest account significantly differs from another prominent theory that is future-oriented, namely Don Marquis' famous argument for the immorality of abortion.

According to Marquis, it is wrong to kill a normal human adult because the victim is deprived of his or her future. The loss of one's life entails the loss of "all the experiences, activities, projects, and enjoyments that would otherwise have constituted one's future" (Marquis 1989: 189). There is no greater loss that can befall a person, and it is the loss of one's future that makes killing prima facie seriously wrong.[4]

Marquis argues that abortion is prima facie seriously wrong for the same reason. "The future of a standard fetus includes a set of experiences, projects, activities and such which are identical with the futures of adult

[4] Marquis's account can admit scenarios involving self-defense and other factors that would make killing acceptable.

human beings and are identical with the futures of young children," Marquis writes. "Since the reason that is sufficient to explain why it is wrong to kill human beings after the time of birth is a reason that also applies to fetuses, it follows that abortion is prima facie seriously morally wrong" (1989: 192). What makes abortion wrong on Marquis's approach is that the fetus is deprived of a future like ours. It suffers the loss of all its future goods, activities, values and projects, and so it too is harmed by death.

This position obviously has something in common with the time-relative interest account. Both view death as a deprivation for entities that experience it. The deprivation for the entity in question has to do with a loss of its future experiences. There is however an important difference between the two accounts, in that the TRI view, unlike the future-like-ours approach, also emphasizes psychological continuity. Aborting a zygote or early-term fetus is not wrong on the TRI account, as there is insufficient psychological continuity between the embryonic human life and the person it will later become to make killing it a deprivation. Because the future like ours approach does not stress psychological unity in the same way, it comes to a different and more restrictive judgment regarding the ethics of abortion.[5]

The fact that the future-like-ours account does not emphasize psychological continuity makes it not merely very different from the TRI account, but much less plausible. As already mentioned, the notion that killing a late term fetus is worse than killing an adult is a difficult notion to accept. Yet this would appear to be an outcome of Marquis's theory, and so a strike against it.

This is not the only problem. An issue for the future-like-ours account concerns the morality of contraception. Marquis believes that his account would face a serious problem were it to entail that contraception is wrong. I think he is right about this. The notion that contraception is ethically wrong is, if anything, even more controversial than the notion that abortion is immoral. Moreover, a position that entails that contraception is wrong for reasons having to do with the loss of a future like ours will need to explain why failing to have as many offspring as possible is not also wrong. After all, we could potentially bring many people into existence, each of whom would have a future like ours. If Marquis' theory entails that

[5] Marquis is deliberately agnostic as to what his theory entails regarding the killing of animals. They too would be harmed by death if they had futures sufficiently similar to ours. However, Marquis does not attempt to outline which animals, if any, have futures similar to ours in this way (1989: 191).

we are obliged to eschew contraception and/or positively obliged to create as many offspring with futures like ours as we can, it will have considerable unwanted baggage.

Marquis is at pains to show that his theory does not in fact rule out contraception. On his view, the theory would have such an outcome "only if something were denied a human future of value by contraception" (1989: 201). According to Marquis however, there is no entity that is denied such a future by birth control. As evidence for this claim, Marquis notes four possible candidates for entities harmed by contraception:

> Some sperm or other.
> Some ovum or other.
> A sperm and an ovum separately.
> A sperm and an ovum together.

Marquis argues that it would be arbitrary to assign the harm to a sperm, "for no reason can be given for making a sperm the subject of a harm rather than an ovum" (1989: 201). It would be just as arbitrary to assign the harm to an ovum rather than a sperm. To say that both are separately harmed by being deprived of futures like ours will not do either, for this entails that two futures are lost rather than one, which is one lost future too many. Moreover, before conception has happened, there are hundreds of millions of sperm that could potentially inseminate the ovum. It therefore seems bizarre to ascribe the loss of a future to a merely possible combination. Hence Marquis's conclusion that there is "no nonarbitrarily identifiable subject of the loss in the case of contraception," while there is such an identifiable subject after conception (202).

Here however Marquis faces a problem. To see it, we need to ask why it would be arbitrary to say a sperm or ovum is deprived of a future like ours (for simplicity, I will focus on the ovum). Marquis's reasoning appears to be that the ovum cannot result in a person on its own. It is rather a necessary condition of a person existing that the ovum be fertilized by a sperm. Insofar as that condition has not yet been met, it is arbitrary to say that the ovum has a future like ours.

But if this is the ground for not assigning harm to the ovum before conception, it does not afford a reason to treat a *fertilized* ovum any differently. For after fertilization there are many other conditions that must also obtain before a person will result. As McMahan writes, "It is obvious that even the normal fetus's intrinsic potential to become a person requires considerable external input for its realization. In the absence of physical protection, nutrition, and, later, exposure to language and culture, and so on, even a normal human fetus will not become a person" (McMahan 2002: 315).

McMahan's remark highlights the fact that conception is only one necessary condition among many of becoming a person. For example, suppose an ovum became fertilized but the woman carrying the embryo then stops eating. A biologically necessary condition of a normal child being born would not have been met. The same would be the case if the woman engaged in activity that was potentially traumatic for her fetus, such as, for example, setting herself on fire, or ingesting substances that cause a miscarriage or severe fetal trauma. A necessary condition of the fertilized egg eventually having a life like ours is that none of these things happen.

Once we recognize this however, it calls into question Marquis' grounds for saying it is arbitrary to ascribe harm to a pre-conception ovum but not a fertilized one (i.e., a zygote). For once an ovum is fertilized, it has merely one less condition to meet on the road to normal existence. But there are further conditions, biological and otherwise, that still must be met. Marquis' argument could thus be reformulated along the following lines to show that it is arbitrary to assign harm to a zygote, that is, an ovum that has been fertilized, but has not yet developed consciousness. For in the stage immediately following conception the candidates for harm are:

Some zygote or other.
Some nutrition source or other.
A zygote and a nutrition source separately.
A zygote and a nutrition source together.

If the reasons Marquis gives for not ascribing harm to an ovum before conception are sound, then the same logic will entail that a zygote is not harmed by abortion. Alternatively, if a zygote is harmed by abortion, as Marquis wants to maintain, then the same will be true of an unfertilized ovum. Either way, Marquis' theory does not in fact show abortion is wrong while permitting contraception.

An important merit of the TRI account is that it does not run into these kinds of difficulties. It does not seek to show abortion is normally immoral, and so does not run into the boundary problem Marquis faces in trying to proscribe abortion while permitting contraception. This is because the TRI account offers a more plausible account of when killing a human entity becomes harmful. On the TRI approach, death is not a harm until the fetus reaches the late-term stage, after it has developed to the point at which can have some minimal degree of psychological unity with its later self. At the zygote stage, when the fetus is not yet sentient, no harm occurs when its life is terminated. In this and other ways, the TRI account's emphasis on psychological unity makes it very different from the future-like-ours account.

Time-Relative Interests and Killing Animals

We are now in a position to apply the notion of a time-relative interest to the issue of killing animals. One obvious implication of the time-relative interest account is that killing a merely sentient animal is significantly less wrong than killing a person. This is because both of the variables that influence time-relative interest, future good and psychological unity, have far less purchase in the life of a merely sentient animal than in the case of a person.

To start with the first, it is normal to ascribe higher value to the goods that are uniquely available to persons. We typically ascribe high value to our ability to reflect on our own lives, form deep and complex emotional bonds, engage in artistic appreciation, scientific discovery, and many other activities that are only available to beings with advanced cognitive capacities. The absence of these capacities means that an entity that lacks them suffers less of a deprivation in being killed, for it is deprived of a more limited quality and quantity of goods, goods to which we invariably ascribe significance in judging the meaning and value of our own lives.

The second variable informing an animal's time-relative interest in its future life, its psychological unity, is also severely limited in the lives of most animals. They have few mental states that refer to previous or later mental states, and the absence of more than minimal memories of the past or desires about the future robs them of one of the major sources of psychological continuity. Unlike human beings, they cannot form complex, detailed beliefs that touch on the span of their lives as a whole. In the life of an animal, McMahan notes, "each day is merely more of the same."

> As an animal continues to live, goods may continue to accumulate in sequence, but the effect is merely additive. . . . There is, in the life of an animal, very little psychological architecture to be carried forward, and earlier and later mental states seldom refer to one another. There is, in short, very little psychological unity within the lives of most animals. (2002: 198)

On such an account, the killing of an animal is not merely less wrong, but *much* less wrong, than killing a person. Indeed, killing an animal, all else being equal, will be less wrong than almost any other form of killing, as can be seen by considering where animal killing fits in on McMahan's spectrum of killing, ranked according to the time-relative interests of the victim.

On the time-relative account, the worst killing-related wrong occurs when a person is killed, because a person not only has a strong good in prospect that is lost, but also exhibits strong psychological unity with the future self that is denied that future good. At the other end of the spectrum is the "wrong" of non-conception, which is not a wrong at all. While it involves a loss of future good, there never exists a being that is denied that future prospective good.

In between these two cases is the killing of an infant (or late-term fetus). It has a significant future good that is lost, but has only a weak prudential unity relation with the future self that is denied that good. This makes infanticide and late-term abortion less wrong than killing a person, but still a wrong above non-conception. Lower even still than killing an infant or late-term fetus is the killing of an animal (keeping in mind our current assumption that psychological unity within a typical animal is very low). For the animal has comparatively far less future good in prospect, and is only weakly related to itself in the future, due to its minimal psychological unity. Hence the time-relative interest account of killing can "endorse and explain the common conviction that infanticide (and late-term abortion as well) is morally a more serious matter than the killing of an adult animal, if other things are equal" (2002: 199).

In the many ways noted so far, the TRI account matches our considered moral views, which is an additional sign of the theory's prima facie plausibility. It bears noting however that the TRI account is at odds with one common view regarding animals. This is the view, already explored at length, that while animals have a legitimate interest in avoiding suffering, they have no legitimate interest in not being killed.[6]

The TRI account of killing cuts against this idea. It holds instead that an animal does have an interest, albeit a comparatively limited one, in not being killed. That means that the interests that are advanced by the animal's death matter in assessing the ethics of killing it. As McMahan

[6] McMahan notes an additional counterintuitive outcome regarding this view, beyond those already mentioned. It would seem to suggest that, all else equal, it always best to euthanize any animal we come across. For in so doing, we would deprive it of nothing of value, taking away a life that counts for zero, while guaranteeing something of disvalue, suffering, never occurs (McMahan 2002: 201). I discuss this outcome in regard to an argument of Peter Singer's in Chapter 6.

points out, killing an animal to eat it when other food sources are widely available is not justified. For the interest that is advanced in this case is not one that is really necessary. If we were on a deserted island and there was only a cow to eat, then killing it would be justified. But the institutionalized system of killing represented by modern agriculture is morally wrong insofar as people in the industrialized world have access to alternative food sources. The many billions of animal deaths that occur to bring us hamburgers, steaks, and bacon represent a moral wrong. If it is not on the scale of the extermination of an equal number of human infants, let alone persons, it is a serious moral wrong in its own right, one a time-relative interest account of the wrongness of killing would counsel us to avoid.

One possible reply to this application of the TRI view to the question of killing animals for food is to emphasize the preference some people have to eat them. Things go better for us when our preferences are satisfied, this reply holds, and so we have an interest in satisfying the preference for meat. If an animal's interest in its continued existence is not as strong as even that of infants, as the TRI account concedes, who is to say that our interest in biting into their flesh does not outweigh the animals' interest in continued life?

This reply appeals to an idea endorsed by preference utilitarianism. It is the idea that moral weight should be granted to the satisfaction of all preferences, without regard to the nature of the preference at hand. As such, the reply shares a widely noted defect of preference utilitarianism, namely its inadequate commitment to equality (Dworkin 1977: 234; 2000: 21–42; Kymlicka 1990a: 35–44). In Chapter 1 we saw that equality is best understood as involving the equal consideration of interests. If we are seriously committed to this view, then we cannot accept that satisfying any preference, regardless of its content, is morally valuable.

This claim can be illustrated by imagining an affluent society in which everyone has an equal share of resources. Within this egalitarian society, there is a city block where everyone has a house and a lawn. All but one of the people living on the block plant a garden. They then turn around and demand that the last person's lawn be kept open for collective use as a park, even though the last block resident wants to plant her own garden.

If our primary goal is satisfying overall preferences, then it is not clear why the people who already have gardens should not go ahead and deprive their neighbor of her own garden. Insofar as she has a preference as to how her lawn is used, she is only one person among many with a preference on the issue. Such would seem to be the conclusion entailed by preference utilitarianism.

It is possible, however, to respond to the demand of the majority in a different way. This is to say that their preference is selfish. What makes it selfish is that it does not take equality into account at its formation. Each member of the majority is disregarding the needs of the last homeowner. As noted by Will Kymlicka, from whom I have borrowed this example, it is hard to see why we should grant moral importance to preferences of this kind. Rather the scenario would seem to highlight a flaw in the idea that preference satisfaction is intrinsically morally valuable. As Kymlicka puts it:

> Utilitarianism [interprets] equal consideration in terms of the aggregation of pre-existing preferences, whatever they are for, even if they invade the rights or commitments of others. But our intuitions tell us that equality should enter into the very formation of our preferences. Part of what it means to show equal consideration for others is taking into account what rightfully belongs to them in deciding on one's goals in life. Hence prejudiced and selfish prejudices are excluded from the start, for they already reflect a failure to show equal consideration. (Kymlicka 1990a: 42)

On such a view, we have no obligation to take preference satisfaction as fundamental. Rather equal consideration comes first, and preferences that do not respect it can be disregarded as illegitimate. Applied to animals, a desire to eat them even when other alternatives are available does not take into account what rightfully belongs to the animal, in this case, its very body. Such a desire fails to recognize that the animal's interest in its own life, while weaker than that of an infant, is stronger than our interest in eating it outside of scenarios in which we would otherwise starve. Such a preference cannot outweigh the animal's legitimate moral interest in continued life, because such a preference fails to amount to a legitimate moral interest at all.

For the same reason, selfish-preferences cannot be aggregated and said to collectively outweigh the animal's interest in its life: in scenarios where non-animal foods are available, it makes no difference if large numbers of people desire to eat a particular animal. All such preferences are ruled inadmissible in our moral calculus for the same reason: they fail to take equal consideration into account at the level of their formation.

Overall, the TRI account of killing has several features that recommend it. One is it can accommodate the intuition that it can be legitimate to cause an animal suffering in order to allow it to keep living. If I hold down a dog and give it a needle, or perform some other operation that causes it temporary pain but saves its life, intuitively it seems I am helping the dog. If we grant that the dog has a legitimate interest in its own future, this intuition can be accommodated. But if that is the case, it gives rise to the corresponding judgment that something goes worse, for the dog, when it is killed, and that this loss requires some form of justification (McMahan 2008: 2).

An additional reason to embrace TRI is that it corresponds with the wide variety of our considered moral judgments McMahan mentions, intuitions regarding The Cure, abortion, non-conception, and similar issues. When the TRI account is criticized in the name of an alternative view of well-being, it will be reasonable to ask how that rival view answers similar questions to those concerning The Cure, abortion, non-conception, etc. For even if it turns out that there are problems with the TRI account that I have overlooked, the fact that it is consistent with so many moral judgments regarding these and other moral issues will still count as evidence in the theory's favor. If some alternative account cuts against more overall considered judgments in these areas than the TRI approach, or commits us to judgments regarding abortion and other issues we consider unacceptable, then that will be grounds to reject the alternative and hang on to TRI, despite its imperfections.

Interest Actualism

This concludes my positive case for the time-relative interest account of the ethics of killing. I want now to turn to a possible objection to my view. This objection is one that charges the TRI account with endorsing a deeply implausible notion known as interest actualism. It holds that the only interests that matter to the moral analysis of killing are those which an individual actually has at the time of death, as opposed to future interests they would have had, had they not died.

This charge against the TRI account has been made by Steven Luper. He begins his case with a modified version of the Cure scenario:

> *Post-Cure Poisoning:* While you were incapacitated, concerned relatives authorized your physician to Cure you, and you were given the treatment on Saturday. The effects on your memory and personality will occur next Tuesday, after which point your life as a profoundly different but happy person is expected to begin. On Monday, however . . . I poison you in your sleep. In the immediate future, the poison I use will cause you to have fewer

> headaches, and no other perceptible symptoms, but later, after your personality is completely transformed, it will kill you. (Luper 2009: 209–10)

Are you harmed by Luper's poison? The TRI account says no – you have no time-relative interests at stake because there is no psychological continuity between you on Saturday and you on Tuesday and beyond. According to Luper, such a conclusion cannot be right.

The Post-Cure Poisoning case is offered to show that the TRI account of the *harm* of killing is wrong. Luper further argues that the TRI theory's account of the *wrongness* of killing is incorrect. To that end he puts forward a second scenario:

> *Poisoned Fetus*: I give the fetus you carry a poison that will have no noticeable effect until the fetus develops into a three year old child or an adult, at which time the poison will be fatal. (Luper 2009: 210)

In this scenario the poison ends the life of a 3-year-old, which is both a serious harm and a serious wrong. But the poisoning has little effect on the time-relative interests of the fetus. This is why the TRI view permits abortion: unlike the case of the Poisoned Fetus, there is no 3-year-old or other entity with a strong time-relative interest in continued life who even comes into being when the fetus is aborted. As McMahan puts it, "Abortion cannot be contrary to the fetus's future interests because it effectively prevents them from ever arising" (McMahan 2002: 284). Luper however asks how the lost life of the child can be wrong in the poisoned fetus case, but in the case of abortion, the life lost by killing the fetus does not count as equally wrong.

McMahan's defense of abortion appeals to the fact that the aborted fetus cannot be harmed by being deprived of interests that never materialize. This raises the possibility that on the TRI account, death is only harmful insofar as it frustrates interests we actually have at the time of death, not interests that would have developed had we been allowed to go on living instead of dying. This is the idea that Luper terms interest actualism. On Luper's articulation, it holds that "our welfare is not affected by what advances (impedes) interests we will not actually have but would have had if we had existed" (Luper 2009: 212).

Interest actualism is a problematic notion. It is at odds with the widespread view that one of the things death deprives us of is the satisfaction of precisely those interests that would have come into being had we not died. If I am grieving the death of a loved one, it is no consolation to be told that my relative cannot be deprived of anything when he is dead, because he no longer has any actually existing interests. It rather feels basic to how we think about death that it deprives the person of future interests. *This* idea is explicitly endorsed on the TRI account, and so interest actualism presents the theory with the additional problem of inconsistency, as the TRI account is now said to both affirm and deny that death frustrates interests not yet actual.

Luper is right to warn us away from actualism. But it is not clear that the TRI account suffers from the problems he identifies. In the case of Post-Cure Poisoning, someone is certainly harmed by being poisoned, but it isn't quite the same someone who takes the cure. It is the different person who exists after the cure. In this way, the TRI account can accommodate the idea that Post-Cure Poisoning harms its victim. What it rejects is the claim that this person is identical to the one who took the cure. On the TRI view, *that* person no longer exists, as he or she has been replaced by someone with the same body, yet a different identity.

Luper's charge that the TRI account appeals to actualism is also not quite right. The matter is complex, however, and showing why this is the case requires recalling the role of psychological connectedness in the TRI account. As we saw, this connectedness is what tethers an individual's future interests to him or her in the present, and hence generates a strong prudential unity relation through time. The fetus, unlike a person, has little such connectedness tying it to the person it has the potential to later become. For this reason the TRI account treats fetuses and persons differently, viewing the death of a person a serious wrong, while the death of a fetus is a minimal loss, one which is easily outweighed by the interests of a pregnant woman who does not want to carry a fetus to term.

Once we recognize the different ways the TRI account treats fetuses and persons, it should become clear that the theory does not endorse actualism. Actualism makes no distinction between the death of fetuses and persons. In either case, it says the only interests that matter are those an individual has at the time of death: future interests are always disregarded. What makes actualism problematic is that it is a wildly implausible way of evaluating the death of *persons*. The passages that Luper cites as evidence of McMahan's alleged actualism, by contrast, have to do with fetuses. A charitable reading of those passages suggests that McMahan is not using the unique case of the fetus to generate principles that should be applied to the death of other beings, up to and including persons. Rather he is seeking to elucidate principles that apply to the specific case of fetal death in a manner that is consistent with the TRI account as a whole. Thus when McMahan notes that abortion effectively prevents the fetus's future interests from arising, he is not outlining a view that applies to any possible manner of preventing future interests from arising – such as killing a person – but referring to a way of preventing future interests from arising that is particular to abortion. And what is particular to abortion is that it prevents the existence of future interests by preventing the existence of a subject who is deeply connect to those future interests through psychological continuity. Abortion and homicide both prevent future interests from forming, but despite this similarity they differ radically in the nature of the victim they involve. One has strong time-relative interests in living while the other does not.

Given this aspect of the TRI account, a better name for the idea it endorses would be *subject actualism*. Such a view holds that in order for death to be a serious harm, it must involve a certain kind of subject. That subject must be one who has a more than microscopic degree of psychological continuity with its future self. This an attribute we have, as do animals to a more limited degree, but fetuses as such do not. To abort an early-stage thus is to prevent its future interests to exist by preventing it from ever developing psychological continuity with an interest-bearing future self. Death does not merely deprive it of a certain set of future interests – it prevents it from developing into an entity that is capable of having serious time-relevant interests of any kind.

If interest actualism is implausible, subject actualism is not. Every account of the ethics of killing will need to distinguish between acceptable and unacceptable ways of preventing a future interest from existing. Consider a view that did not draw such a distinction, and said that preventing future interests from coming into existence was always wrong, regardless of the method. Such a view would entail that non-conception – failing to conceive a child when it was within our power to do so – was immoral. Such a view is extremely difficult to accept.

Abortion involves a prevention of future interests in between non-conception and killing a person. The TRI account, rather than implausibly endorse actualism, makes the more reasonable move of positing a threshold of psychological continuity that a subject must possess before its death becomes seriously wrong. In positing such a threshold, it outlines a view of the ethics of killing that is sensitive to the particular status of fetuses, which is something any minimally adequate theory must do.

The TRI Account and Life-Saving Infanticide

I have mentioned the various ways in which the TRI account does and does not accord with familiar intuitions. There is a final area where McMahan's presentation of the theory is very much at odds with everyday morality. This concerns this issue of life-saving infanticide.

McMahan asks us to imagine a woman who wants to be a single parent and becomes pregnant via artificial insemination, only to die during childbirth. She has no friends and no family. When her baby is born it is a healthy ordinary child. However, in the same hospital are three 5-year-old -children who will die unless they soon receive organ transplants. Imagine it turns out that the newborn is exactly the right tissue type. According to McMahan:

> It ought to be permissible, if other things are equal, to sacrifice the newborn orphaned infant in order to save the other three children . . . [because] its time-relative interest in continuing to live is much weaker than that of each of the three older children; it is only one individual where they are three; and it has no parents or relatives, while each of them, we may suppose, is cherished by its parents and others. (2002: 360)

As McMahan goes on to note, "most people will find this implication intolerable, and I confess that I cannot embrace it without significant misgivings and considerable unease" (2002: 360). But on a TRI account, he argues, there is no way to avoid this judgment. The TRI approach can accommodate the notion that it would be wrong to make a similar sacrifice of a person. Persons have a much stronger interest in their future, which can ground the "deontological constraint" of a right to life that preclude killing them to save three other patients (2002: 358). But infants, like fetuses, have a much weaker time-relative interest in their future, and so fall below this threshold of moral respect. McMahan concludes that this permits trading off their interests against that of a larger group, in this case, the three children.

I disagree with McMahan that tolerating the sacrifice of infants is a necessary upshot of the TRI view. An argument that I give later on, concerning the doctrine of double effect, provides reasons to doubt this view. Because my grounds for philosophical disagreement with McMahan arise out of this later argument, I will hold off presenting them until after outlining the arguments in question. In the meantime however it bears noting that even if McMahan were right on a purely ethical level, there would still be reason to endorse a legal ban on infanticide.[7]

One reason is because the circumstances in McMahan's though experiment will rarely apply. The number of infants without any family whatsoever is small. McMahan illustrates his case with reference to a woman dying in childbirth, an outcome that is now rare in the industrialized world. In the U.S., which sees such deaths more often than other wealthy countries, such deaths occur 16.7 times out of every 100,000 births (Rogers 2010). The number of times such deaths involve cases in which there is no father or extended family with an interest in the child will be vanishingly small. A law against life-saving infanticide will protect all infants while running little risk of preventing genuinely justified killings.

Permitting life-saving infanticide also risks outcomes in which, rather than save three lives, something goes wrong – and the three children as well as the infant *all* die. A ban on infanticide avoids that risk. Such a ban also avoids any possibility of the procedure ever being performed on the wrong child, a possibility that could give rise to the separate problem of expectant mothers failing to be reassured by being told that infanticide is only a policy in cases involving infants unlike their own, and preferring not to give birth in any hospitals that kill infants. In short, one need not be a consequentialist to suspect that overall consequences would be best if the practice were subject to a complete institutional ban.[8]

These considerations are of course legal rather than ethical. They provide no purely ethical grounds for objecting to life-saving infanticide in thought experiments one might devise involving remote islands where there is only a surgical team, the infant and the three children; cases in

[7] Such a ban would not prevent ending the life of an infant born with anencephaly, a rare condition in which infants are born without a functioning cerebrum. Infants with this condition are not conscious, do not feel pain and commonly die within hours or days of birth. When anencephalic infants are allowed to live out the short course of their natural lives, the blood flow to their organs is commonly so reduced that it renders their organs unfit for transplantation. The American Medical Association briefly endorsed removing the organs of such infants (with parental permission) prior to complete brain death, before reversing its endorsement in the face of protest (McMahan 2002: 208). Insofar as such infants will soon die of natural causes, and are alive only in the minimal sense of being a functioning organism with not capacity for consciousness, the considerations for an infanticide ban do not extend to them.

[8] These considerations also apply to proposed infanticide cases in which the infant in question is mentally handicapped, which McMahan also discusses, but not to cases of euthanasia (2002: 358). Among other differences with life-saving infanticide, euthanasia is done for the benefit of the person being killed, and he or she is often able to consent to it.

which life-saving infanticide is performed in secret; or other scenarios in which the aforementioned considerations do not apply. Some readers may find even these possibilities disturbing enough. But it seems worth noting that even if McMahan's view of the ethics of infanticide were correct, and even if the more philosophical objections that I will later present fail, it would still be appropriate to legally outlaw the practice, thereby forestalling the objection that the TRI account would usher in widespread barbarism if it were accepted.

CHAPTER 3

Burger Veganism

Introduction

I turn now to new omnivorism, which finds one of its major proponents in Steven Davis (2003). Davis has offered an influential argument for the view that the diet most consistent with animal protection philosophy is one that contains some meat. The crux of his account has to do with the accidental death of field animals during vegetable harvesting. Empirical studies suggest that the number of field mice, rats, and similar creatures killed in crop cultivation may outnumber the total animal deaths involved in the raising of beef cattle, so long as the cows are raised on a diet of grass rather than grain. If so, then the most logical diet for animal advocates to adopt is one that includes hamburger and milk from grass-fed cows, in order to reduce the overall number of animals killed.

This version of new omnivorism – which we can call the burger vegan view, for reasons explained below – has attracted noteworthy media attention, including favorable mentions in cover stories in *Time* and *The New York Times Magazine*, and variations on Davis's proposal involving beef and other meats have subsequently appeared in the philosophical literature (Corliss 2002; Pollan 2002; Schedler 2005; Cahoone 2009; Bruckner 2016; Demetriou and Fischer 2018). No doubt much of the attention is due to the novelty of Davis's position. He offers an immanent critique of animal protection theory in general and the philosophy of Tom Regan in particular. Davis's argument does not challenge core tenants of protection theory outlined in previous chapters such as anti-speciesism, equal consideration or time-relative interests. Instead, Davis's argument accepts Regan's irreproachably protectionist claim that, when it comes to making comparisons between non-human vertebrates, species is irrelevant to moral status: all else equal, killing a mouse is just as wrong as killing a cow. However, unlike Regan, Davis thinks an argument based on this and related premises undermines, rather than supports, the case for a meat-free diet. Thus if Davis is correct, one of the leading theories of animal protection will be hoist by its own petard on the central practical question it engages, namely, what to eat.

Burger veganism is based on an ingenious argument, one that displays an elegant simplicity in the way it turns animal protection theory on its head. Despite its genuinely appealing qualities, however, burger veganism has some notable defects. To start with, the scientific studies on which Davis relies actually document two different forms of harm to field animals: there are those directly killed by harvesting equipment and those that become the prey of other animals. Once this distinction is made explicit, the degree to which such studies pose a problem for animal protection theory considerably weakens. Davis also overlooks philosophically significant forms of harm to human beings that are present in beef production but not vegetable harvesting. Finally, he bases his argument on the implausible assumption that there is no difference between deliberate and accidental killing – either of an animal or a person. These problems ultimately prevent Davis from offering a successful critique of Regan.

Problems with Davis's empirical calculations carry over to other arguments for meat-eating that presume his calculations are correct. Two variations on burger veganism, however, are untouched by criticisms of Davis's estimates. In one case this is because the argument is based on independent data, in the other because it presumes only that some animals are harmed in plant cultivation and not that Davis's calculations are unsound. These variations on burger veganism, which respectively involve the consumption of free-range Australian beef and roadkill, have problems of their own that prevent their arguments from going through. Finally, there is an overlooked issue not only for burger veganism but all of its descendants, in that more than one current trend in plant agriculture causes little or no collateral harm to animals. Taken together, these trends suggest that field animal deaths are a historically contingent problem that in future may be reduced or even eliminated altogether.

The Davis Diet

Some of the power of Davis's paper is illustrated by two vivid passages he quotes from other writers describing the death of field animals. From novelist Barbara Kingsolver: "She stopped speaking when her memory lodged on an old vision from childhood: A raccoon she found just after the hay mower ran it over. She could still see the matted grey fur, the gleaming jaw bone and shock of scattered teeth" (Davis 2003: 388). From hunting advocate Ted Kerasote: "When I inquired about the lives lost on a mechanized farm, I realized what costs we pay at the supermarket. One Oregon farmer told me that half of the cottontail rabbits went into his combine when he cut a wheat field, that virtually all of the small mammals, ground birds, and reptiles were killed when he harvested his crops. Because most of these animals have been seen as expendable, or not seen at all, few scientific studies have been done measuring agriculture's effects on their populations" (Davis 2003: 389). Animal ethicists have sometimes used the image of an expanding circle to express the idea that we should extend our moral concern outward to include animals. Davis essentially replies that if that is the case, then the circle will teem with all the animals of the field that are potential casualties of modern agriculture. As he points out, even a partial list of field animals in the United States would include opossum, rock dove, house sparrow, European starling, black rat, Norway rat, house mouse, chukar, gray partridge, ring-necked pheasant, wild turkey, cottontail rabbit, gray-tailed vole, and many different species of amphibian.

Davis's critique of Regan begins with an estimate of the percentage of field animals killed in crop harvesting. Davis derives his estimate from two scientific studies, the first of which was conducted in England, and examined the effect of wheat and barley farming on mouse populations (Tew and Macdonald 1993). The scientists found that the number of mice per hectare declined from 25 to five after harvesting. As Davis notes, "This decrease was attributed to both migration out of the field and to mortality. [The authors] estimated the mortality rate to be 52%" (2003: 389). The second study, conducted in Hawaii, found that sugarcane production inflicted a 77% mortality rate on Polynesian rats (Nass et al. 1971). Davis concludes that the average number of animals killed by agriculture will be somewhere between 52% and 77%, and settles on 60% as a rough approximation. Assuming that the population density of 25 mice per hectare cited in the English grain paper is typical of field animals in general, Davis writes, then we can calculate the average number of field animals killed in any agricultural operation to be 60% of 25, or 15 animals per hectare per year.

Davis contrasts this mortality rate with that caused by grass-fed beef raised on a pasture-forage model. Pasture foraging requires fewer passes through the field with tractors and farm equipment than does intense crop production. For that reason alone, Davis writes, it is likely to reduce the number of field animals killed. But that number can be reduced even further when grazing cattle are used to harvest the forage, converting it into beef and milk. There are no statistics available regarding exactly how many field animals are killed in pasture-forage cultivation, but Davis says it is reasonable to assume it would be only half the number killed in crop production, or 7.5 dead animals per hectare.

There are 120 million hectares of cropland in the United States. If all of that land were used to harvest plant food, and if 15 animals were killed per hectare, the total number of animals killed every year would be 1.8 billion. Davis calls this arrangement the vegan model. He contrasts it with an agricultural system under which only half of American cropland is devoted to plant food, while the other half supports ruminant cattle raised on a grass diet. We can call this arrangement the Davis model. The question Davis seeks to answer is how many dead field animals his model would result in. He calculates the answer according to the following formula:

60 million hectares, plant production × 15/animals/ha = 0.9 billion
60 million hectares, forage production × 7.5 animals/ha = 0.45 billion
Total: 1.35 billion animals

The goal of Davis's analysis is to determine what diet best upholds Regan's least harm principle, which reads as follows: "Whenever we find ourselves in a situation where all the options at hand will produce some harm to those who are innocent, we must choose that option that will result in the least total sum of harm" (Davis 2003: 388). Regan's argument ascribes no moral significance to species membership, and abides by the equal consideration of interests. This is evident in the fact that his argument ascribes equal moral status to merely sentient beings and bestows persons with a higher status on the grounds that they have interests beyond those of merely sentient beings. Davis's reconstruction of Regan's argument thus assumes the two central protectionist principles. Davis takes his reconstruction to show that Regan should cease his longstanding advocacy of a vegan diet, and join Davis in a nice grass-fed steak (perhaps with some ice cream for desert). For the diet that actually embodies Regan's least harm principle is one cultivated according to the Davis model, as it would kill 450 million fewer animals per year.

When Davis wrote his paper, 8.4 billion food animals were slaughtered in the United States annually. Over 8 billion were chicken, 37 million were cattle, and the remainder were pigs and other species. In an important passage, Davis notes that even if his proposal resulted in a large increase in the number of cattle killed to compensate for the decrease in chicken, it would still be superior to a plant-only system. "Even if the numbers of cows and calves killed for food each year was doubled to 74 million to replace the 8 billion poultry, the total number of animals that would need to be killed under this alternative method would still be only 1.424 billion, still clearly less than in the vegan model" (ibid., 391).

Davis argues that his model is superior to several other proposals that have been put forward. People for the Ethical Treatment of Animals, for example, has made the (surely facetious) suggestion that if we continue to eat meat, we should consume the largest animals possible – blue whales – in order to reduce the overall number of food animals deaths. Davis says this would be impractical, as it would be impossible to kill enough whales without driving the species to extinction. A different version of the largest-animal-possible proposal would see people dining on elephants and draft horses, which can grow to twice the size of cows. Davis argues that this is unlikely to happen, however. Otherwise, there would already be more people willing to eat horses.

This, then, is Davis's argument. Before going on to critically assess it, it is worth recalling that Davis puts his pasture-ruminant model forward as an alternative to veganism. Yet whether this is true depends on how we understand veganism. Many common definitions do characterize it according to a list of particular prohibitions involving food and other animal products. A more plausible definition, however, characterizes it as a commitment to the principle of least harm itself. According to the advocacy group Vegan Outreach, for example, veganism "is about lessening suffering and working for animal liberation as efficiently as possible" (Vegan Outreach 2005). As one of the organization's founders puts it, "the issue . . . isn't, 'Is this [food] vegan?' Rather, the important question is: 'Which choice leads to less suffering?' Our guide shouldn't be an endless list of ingredients, but rather doing our absolute best to stop cruelty to animals" (Ball 2005). Vegan Outreach is concerned with animal suffering, but veganism can be taken to also be concerned with reducing the unnecessary killing of animals. Vegans disagree among themselves as to whether some foods, such as honey, are vegan or not.[1] Among the benefits of defining veganism in terms of harm reduction is that it treats disagreements

[1] The debate over honey has to do with whether or not honeybees warrant moral respect. Most animal philosophers exclude insects from the realm of moral concern, on the grounds that they are not vertebrates and do not have a central nervous system or nociceptors (pain receptors) in their brains. See Varner (1998) for a summary of six studies which collectively suggest that insects, unlike vertebrates, do not feel pain. (For an online summary of Varner's research, see http://philosophy.tamu.edu/~gary/intro/pain.html). Some vegans, however, argue that simple conscious awareness, rather than suffering, is the foundation of moral respect, and so eschew products made from insects. I assume this second group represents only a tiny proportion of vegans, but am unaware of any survey data.

over any particular food as secondary, and emphasizes instead the underlying belief ethical vegans have in common.

On this view, the Davis diet is itself a variant of veganism. I believe this is a more accurate way to describe it than Davis's own characterization. Davis, it should be noted, does not challenge the principle of harm reduction, merely its prohibition against consuming beef and dairy products produced from grass-fed cattle. (Although he does not mention it, his proposal could also justify wearing leather, which vegans normally object to). Changing one's lifestyle due to a commitment to reducing animal suffering would entail rejecting many practices and institutions that are now common, including wearing fur, hunting, visiting roadside and perhaps other zoos, cockfighting, the seal hunt, trapping – in short, the familiar list of things animal protectionists condemn. As for eating habits themselves, it is worth noting that only 3.3% of Americans are vegetarian and of those only half, or roughly 1.6% of the total population, are vegan (Vegetarian Resource Group 2016). Davis's rejection of all poultry-related products places him among the less than 2% of society that objects to eating eggs on ethical grounds. Characterizations of Davis's disagreement with Regan that frame it as a choice between meat-eating and vegetarianism are therefore not accurate.[2] At stake in the debate are really two different diets that seek to radically reduce the amount of harm inflicted on animals – in short, two different versions of veganism.

The Wheat and Sugarcane Studies

Davis's discussion of field animal deaths highlights a previously overlooked type of animal harm, one that, to my knowledge, no previous animal protection theorist has identified. For that reason alone, animal advocates are in his debt. Reading Davis's highly original paper one has the immediate sense that it contains an important argument that must be grappled with. Indeed, I would go so far as to say that Davis has offered the most thought-provoking response to animal protection philosophy since the ethical debate over animals began in earnest 30 years ago.

But is Davis's argument persuasive? Looking at it carefully reveals it to have problems, beginning with the conclusions Davis draws from the empirical studies he cites. Start with the English mouse study. The researchers actually found that the number of animals killed directly

[2] See the previously-mentioned media stories for characterizations of this kind. Also note the title of Matheny (2003).

by farmers at harvest time was fairly small. One of their experiments, for example, examined the survival rates of 33 radio-collared mice before and after harvest. Only one mouse was killed by the combine harvester, a death rate of 3%. After the harvest, the scientists went on to track eight mice during hay bailing. This phase of cultivation killed zero mice. A third experiment involved stubble burning, when the remaining hay stalks are set alight to clear the field. Here the researchers did find a significant death rate, affecting two out of five mice they tracked, or 40%.[3] But not only is that figure lower than the 52% mortality rate Davis derives from the study, but the researchers themselves note that it is difficult to draw any conclusions from an experiment with such a small sample size. It should also be noted that burning is not an inevitable aspect of wheat or barley cultivation. The stubble can be plowed back into the field or eliminated in the same manner as Davis's paper highlights, namely, by allowing ruminant animals to graze on it. Given these considerations, it is perhaps unsurprising that Davis does not include the stubble burning experiment in his calculations.

So where does the 52% figure come from? It actually concerns the number of animals that were killed by other animals after the crop was removed. One such animal the researchers observed were tawny owls. As the authors note, harvesting the field deprived the mice of their protective cover, making them more exposed to their natural predators:

> [O]wls changed their hunting behaviour at harvest. Prior to harvest, they rarely hunt farmland. When they do, they concentrate particularly on the grassy banks bordering hedgerows since the crop prevents them from stooping on prey. Immediately following harvest, however, the owls fly low over the fields and stoop onto prey both in the stubble and directly into the straw lines left by the combine. (Tew and MacDonald 1993: 282)

The effect of owls and weasels, another local predator, was far more dramatic than the farm machinery. Seventeen of 32 radio-collared mice disappeared in the week after harvesting. In nine cases the scientists actually saw the mouse killed or came upon evidence, such as chewed collars in weasel nests, after the fact. While it is impossible to rule out the possibility that some of the other missing mice simply fled the field, the

[3] The researchers note that whether or not the mice were killed by the fire depended on the location of their burrow: "Both mice that were killed by stubble burning had burrows directly below the straw lines, where the stubble fire burnt most fiercely. All three survivors had burrows between the straw lines. Post-mortem examination suggested that the mice had been asphyxiated by the stubble fire, and it is likely that as the fire passed overhead it drew the air out of the burrow system" (Tew and Macdonald: 1993: 282).

researchers were monitoring for this possibility and conclude that it is more reasonable to assume that all 17 disappeared rodents were taken by predators.

The fact that these deaths were caused by animals rather than human beings poses a problem for Davis's argument. The animal protection movement as a whole has long drawn an ethical distinction between animals killed by people versus those killed by animal predators (Singer 1990: 226; Fox 1999: 163). Tom Regan is no exception. As he writes, "in claiming that we have a prima facie duty to assist those animals whose rights are violated, therefore, we are not claiming that we have a duty to assist the sheep against the attack of the wolf" (Regan 2004: 285).

Animal philosophers offer different rationales for the predation exception. Singer, for example, makes a practical argument, saying that it would be impossible to police nature, so the issue is moot. Michael Allen Fox agrees with Singer, but also argues that saving mice is the same as depriving owls of their fulfillment, and so itself involves a form of harm. Regan, for his part, bases the distinction on the fact that predators cannot engage in moral reasoning. Wolves and owls are not themselves moral agents, so they cannot be said to have moral duties. The obligation to avoid directly harming animals is therefore one that only human beings must uphold. In the case of the owls hunting the mice, on Regan's view it is irrelevant that a human action, cutting down the barley, allows the owls to better see their prey. The mere fact that the mouse is harmed by a non-human animal is enough, according to Regan's philosophy, to make it morally unobjectionable.

The fact that Regan makes an exception for predation brings Davis to a fork in the road. On the one hand, Davis could conceivably reject the predation exception. Simply in virtue of the fact that Regan offers a theory of animal protection, Davis might hypothetically argue, means that Regan is obliged to condemn both human harm to animals and predation. Were Davis to take this line of attack, however, he would only create problems for himself. The charge that an ethical concern with non-human suffering entails condemning predation, and so policing nature, is often made by critics of animal protection. Usually it is put forward as a *reductio ad absurdum* to show that animal protection theory is not worth taking seriously. However, it is unclear that Regan or any other animal protectionist is in fact obliged to condemn predation. As Fox points out, many animals are wired to hunt, and preventing them from exercising this capacity may deprive them of something essential to not only their own flourishing, but essential to the continued existence and flourishing of nature itself.[4]

Moreover, in the case of the mouse study, the authors themselves note that although cutting down the crops is a disaster for the mice, it is a bonanza for younger owls and weasels, many of whom would die without the food source harvesting makes available (Tew and Macdonald 1993: 282). All things considered, harvesting the field might even do more good for animals than harm, as it is possible that many of the mice will die soon anyway. As the authors write, "Whilst the mortality of the arable wood mouse population at harvest is dramatic, many of the animals killed at this time of year are nearing their natural life expectancy, and it may be that the predation corresponds to a 'doomed surplus'" (ibid., 282).

Even leaving such considerations aside, however, the nature of Davis's own argument obliges him to draw a distinction between animals killed by human beings and predation. Were Davis to maintain that we are obliged to prevent animal suffering in nature, it would not lend support to his argument concerning field mice run over by harvesters. Rather it would drain his position of any plausibility – the idea that we have an obligation to police nature is considered ridiculous for good reason. Such a stance would also be at odds with Regan's position regarding predation. Davis's critique, however, derives its distinctiveness and force precisely from the fact that it starts from within Regan's own philosophy. Such a project would be rendered pointless by an argument that called one of Regan's major premises into question.

To preserve the original and immanent quality of his argument, Davis needs to take the second fork, to take the path that stays within Regan's premises. Very well, Davis might say, I accept that acts of predation cannot be invoked as part of an immanent critique. So in the case of the English mouse study, the relevant percentage is not 52% predator deaths, but 3% killed directly by the combine harvester. According to Davis's estimate of 25 animals per hectare, that works out to 0.75 deaths per hectare. If we plug that figure back into Davis's model involving 60 million hectares of farmland devoted to raising plant food, the overall death total is not 0.9 billion,

[4] In citing Michael Allen Fox's argument for the predation exception, I am aware that I am not offering a purely immanent defense of Regan. Like Fox I believe the inability of predators to engage in moral reasoning is a relevant consideration, but is not by itself sufficient grounds to justify an exception for predation (For criticism of Regan on this score, see Jamieson [1990]). However, this strikes me a minor deviation from Regan compared to the one I shall criticize in Davis. For a protectionist defense of policing nature under highly counterfactual conditions (and so unrelated to contemporary agriculture), see McMahan (2016a).

but 45 million. That is less than the 74 million cattle deaths Davis allows in his preferred agricultural system, to say nothing of however many field animals are accidentally killed in cattle pastures (let alone all the coyotes and other predators ranchers shoot to protect their herds).[5] That leaves Davis with the sugarcane study.[6] Here he would appear to be on stronger ground. The death rate it found regarding rats did not include deaths due to predators: the 77% mortality rate was entirely from rats killed by harvesting machinery. Davis, however, makes a small mistake by citing this as an annual death rate. Hawaiian sugarcane has a longer growing time than other crops. The wheat and barley in the English mouse study, for example, were harvested once a year. Hawaiian sugarcane has a two-year growing season (the authors of the sugarcane study note it varied from 22 to 38 months in 1970, the year the study was conducted). We therefore need to cut the 77% figure in half (assuming a two-year season) to get the real annual number of 38.5%. That is still a significant mortality rate of 9.6 animals per hectare, larger than Davis's original estimate for grass-fed beef production. So even granting the lower death rate in the English mouse study, Davis might say, his argument still stands.

However, it is important to recognize the differences between the two studies, which describe very different agricultural processes. The authors of the Hawaiian study note that in sugarcane production, the field is first set alight to burn away dead leaves. As the scientists go on to write:

[5] In fairness to Davis, the number of predator animals killed by ranchers is much smaller than the number of food animals. In the U.S. during 2002 they amounted to 86,000 coyotes, 5,000 foxes, 380 black bears and 190 wolves (Marcus 2005: 198). On the other hand, Erik Marcus also mentions a 98% decline in the prairie dog population, which numbered 5 billion at the end of the nineteenth century. This however, may be more of an environmental cost than anything else. On the issue of the environmental cost of animal agriculture, it has become common for animal advocates and environmentalists to claim that it takes 5,200 gallon of water to produce one pound of beef. Marcus convincingly debunks this apocryphal statistic, noting that "the environmental impact of meat production is far less costly than many vegetarian activists suggest" (ibid., 195).

[6] On page 389 of his paper Davis also cites the following sentence fragment: "mowing of alfalfa caused a 50% decline in grey-tailed vole population." The fragment is taken from Edge (2001). However, when one looks up the passage to which Davis is referring, it reads, "*Edge et al. reported that* mowing of alfalfa caused . . . ." In other words, the sentence in question is itself a citation, and the ultimate reference is to Edge et al. (1995). When one turns to this second paper, it turns out to be a study of the effects mowing has on vole birth rates and migration patterns. Importantly, the scientists do not describe any voles being killed by the act of mowing itself. It is also worth noting that the study did not take place in an actual alfalfa field. Rather, it was conducted at an experimental farm where alfalfa was planted inside 45 by 45 meter enclosures, which were "constructed of galvanized sheet metal extending approximately 1 m above ground and 0.6–1.0 m below ground level to prevent escape by burrowing rodents" (ibid., 246). A study with this aim and methodology is not relevant to a discussion of field animals killed by harvesters. Davis would appear to have misunderstood what Edge meant in his single-authored paper by population "decline."

> [T]hen a v-cutter (a caterpillar tractor equipped with a snowplow-like device) cuts the stalks at ground level and leaves them in [rows]. Pushrakes (caterpillar tractors with brush rakes instead of bulldozer blades) push the cane into piles. Self-loading transporters pick up the [rowed] stalks and haul them to a loading point from which they are transferred to trucks and hauled to the mill. (Nass et al. 1971: 355)

This is quite unlike the process the mouse study describes. Indeed, in one respect the harvesting methods actually have opposite outcomes. As we saw, one of the experiments in the mouse study suggested that burning the stubble killed the most animals. The sugarcane study, by contrast, found that the rats always survived the fire. What really did them in were the tractors, which often ran them over (in a few cases, the rats were inadvertently transported by truck to the sugar mill and put through the washer). Importantly, the sugarcane study makes no mention of combine harvesters, just as the mouse study does not refer to tractors.

When the differences between the studies are made explicit, we see what a leap Davis makes in using them to determine an average annual death rate for all agriculture harvested in the continental U.S. (where neither study was conducted). One could just as easily conclude from the same research that every different form of crop harvesting has a different mortality rate for field animals, and that a national average will need to be based on more than two studies. Alternatively, it might be the case that most forms of agricultural production are more like wheat than sugarcane harvesting, or use combines rather than caterpillars. If so, then Davis's analysis would have little relevance beyond the sugarcane sector (which vegans should perhaps hope to see replaced with sugar beets, which grow beneath the soil and already account for over 50% of U.S. sugar production) (USDA 2005).[7] After reading the same studies as Davis, my own conclusion is that the science of estimating field animal deaths is still in its infancy, and is not a good basis on which to make large-scale recommendations. Davis himself concludes that more research is needed in this area (Davis 2003: 393), but we do not know enough to make even the rough calculations that Davis offers.

The central problem here concerns generalizing field animal deaths across species and crops. The death rates of crop harvesting practices do not appear to follow a one-size-fits-all model. Rather, there appears to be wide variation in the effects particular crop harvesting practices have on particular species. This can be seen by noting the findings of additional studies of the effect of harvest on field animals.

[7] See also this chapter's concluding discussion of the end of the Hawaiian sugarcane industry.

A 2004 study examined the effect of wheat and corn harvesting in central Argentina. It compared the population and distribution of grass mice (*Akodon azarae*) in three habitats: crop fields, regions bordering the fields and the wider surrounding area. While the number of mice found in fields substantially decreased after harvest their numbers substantially increased in the border regions. When it came to "disappearances," a category that included both mouse deaths and migration out of the study area, there was no significant difference between the three habitats. The study concluded that changes in the number of field animals were "the consequences of movement and not of high[er] mortality in crops" (Cavia et al. 2004: 98).

A second study tracked the presence of small mammals in and around cornfields in South Dakota. Snap traps set by the researchers caught short-tailed shrews, masked shrews, prairie voles, meadow voles and three species of mice. However, many of the species caught adjacent to cornfields were not found in the cornfields themselves. "With the exception of one grasshopper mouse, deer mice and white-footed mice were the only species captured in cornfields" (Pinkert et al. 2002: 41). The authors suggest that mice are the only permanent residents of corn fields because they don't require herbaceous plant cover to avoid predators, and can rely instead on underground burrows (2002: 43; Warburton and Klimstra 1985: 329; Johnson 1987: 128). The populations of deer and white-footed mice in the cornfields after harvest were unchanged in one of the two crop-field habitats the researchers examined, while in the other the number of both species increased slightly. (The same was true of grasshopper mice, the population of which increased from zero before to one after harvest.) The evidence suggests mice likely migrate into croplands after harvest to eat the seeds and crop waste available there.

The authors of the South Dakota study note that their findings may underreport the presence of shrews because they can avoid capture in snap traps more easily than other traps. Despite this limitation, their study and its Argentinian counterpart are in keeping with two common finding in the literature on field animal populations. One is that field-animal populations are species-specific. A cultivated area that supports large numbers of one local species can be home to zero members of another. This already makes generalizing mortality rates across species difficult. But the second finding is that crop cultivation often has no effect on whether field animals live or die.

Some studies do suggest that no-till agriculture has more positive outcomes than conventional tillage systems, such as larger and more stable population of small mammals, as well as greater species diversity (Warburton and Klimstra 1984; Young 1984; Johnson 1987). Other studies however suggest that the comparative disadvantage of conventional tillage is illusory, as there is "no adverse effect by tillage on resident small mammals" (Fleharty and Navo 1983; Castrale 1985; Getz and Brighty 1986; Albers et al. 1990; Wegner and Merriam 1990; Schwer 2011: 29). What all these studies have in common however is that they don't document animals being killed in crop cultivation. Indeed, it is relatively uncommon to find peer-reviewed research documenting such an outcome.

This absence of evidence poses a problem for any high estimate of a fatality rate caused by harvesting machinery. It is worth underscoring the limits of our current understanding regarding the frequency of these deaths. In addition to the studies cited by Davis, I have only been able to find one further study documenting a field animal's death directly due to crop harvest. Jacob and Hempel (2003) examine four tracts of agricultural land that saw 12 voles die shortly after harvest. One of the voles was killed by farm machinery and another was removed by a buzzard. No cause of death could be identified in the other cases but the authors note that some may have been buried alive in their burrows due to soil compression caused by farm machinery (ibid., 49). The two studies originally cited by Davis documented the deaths of three mice and 17 rats, respectively. When Jacob and Hempel is added, the three studies together document a maximum of 31 field animal deaths due to agricultural machinery. If the Hawaiian rat deaths are removed from the 31 total, that leaves 14 (only four of which were confirmed to be due to agriculture). Such a sample size would seem too small to support any large-scale generalizations about agricultural practices.

Land Fertility and Agricultural Harms to Human Beings

Nonetheless, suppose that we give Davis the benefit of the doubt. Suppose that more studies are done, and they find that Davis's original mortality rate is typical of American agriculture as a whole. Even if that were the case, Davis's analysis would still suffer from two problems that defeat his attempt to offer an immanent critique of Regan and of protection theory more broadly.

The first problem has already been pointed out by Gaverick Matheny. Davis's calculations are based on the assumption that one hectare of land devoted to grazing cattle produces as much food as one hectare devoted to crops. In fact, crop production takes much less land to produce a similar amount of food. Matheny notes that in one year, 1,000 kilograms of protein can be grown in one hectare of cropland planted with soy and corn. It would require 2.6 hectares of pasture for dairy cows fed on grass to produce the same amount of protein, and 10 hectares for grass-fed beef cattle to do so (Matheny 2003: 509).[8] So even if we allow Davis his original estimates that 15 animals die per hectare in crop production, and that only half as many are killed by pasture foraging, the comparative superiority of the Davis model is undone by the fact that it will require far more hectares. As Matheny points out, adults are recommended to have 20 kilograms of protein annually. Given Matheny's revised calculations regarding land fecundity, a vegan would kill 0.3 field animals to obtain that amount, a lacto-vegetarian would kill 0.39, and someone on the Davis diet would kill 1.5 (2003: 507). In other words, if Davis's analysis turned Regan on his head, the effect of Matheny's response is to turn Regan right side up again. As Matheny puts it, "[after] correcting Davis's math, we see that a vegan-vegetarian population would kill the fewest number of wild animals, followed closely by a lacto-vegetarian population" (2003: 507).

Matheny's objection is a powerful and important one. However, the productivity of all forms of agricultural land has increased over time and there is a possibility (however remote) that grazing land may experience a productivity boom at some point in the future (Brown 2003). Matheny's argument would also seem to leave open the question of what to do with land that is not suitable for crop production, yet which could support

[8] Matheny suggests in a personal communication that crop land may in fact be even more productive than he originally estimated. At the outer limit, he writes, land devoted to grazing cattle in a cold climate is 67 times less productive than land devoted to soy crops. Matheny's revised estimate is based on a soybean crop yield of 37 bushels per acre, with an average meal yield of 48 lbs. per bushel, of which 48% is protein (figures taken respectively from (http://www.unitedsoybean.org/soystats2002/statesoystats/cropyield.html; http://www.unitedsoybean.org/soystats2002/miraclecrop.html and http://www.ag.iastate.edu/aginfo/news/1997/soyquality.html). That works out to 852 lbs. protein per acre (37 × 48 × 0.48), or 955 kilograms protein per hectare. Grazing cows, by contrast, gain an average of 131 kilograms per hectare, with an average 0.42 kilogram of meat per kilogram of liveweight, and 0.26 kilogram of protein per kilogram of meat (figures taken, respectively, from http://jas.fass.org/cgi/content/full/82/8/2503/T3; http://ars.sdstate.edu/MeatSci/May99-1.htm and http://www.calorie-count.com/calories/item/23502.html). That works out to 14.3 kilograms of protein per hectare (131 × 0.42 × 0.26), or one 67th the amount produced from soy. As Matheny notes, pasture finished cattle might have a different total efficiency, and grass-fed cows in a hotter climates could be considerably more productive. Nonetheless, he believes his revised figures are more representative than those in "Least Harm." For the purposes of rebutting Davis, the difference between Matheny's 1:10 and 1:67 ratios is academic, so I have continued to use the 1:10 figure in the main text. Given Davis's estimates regarding the number of animals killed in beef and crop farming, anything above a 1:2 ratio invalidates his argument.

ruminant animals. Finally, suppose it turns out there is some particular type of vegetable that involved extremely high field-animal mortality rates, so high that when the harvesting method of the especially bloody vegetable is compared to raising dairy cows, harvesting the vegetable results in a higher number of animal deaths – even after Matheny's corrections are taken into account. If that were the case, Matheny's criticism would not apply to dairy products from grass-fed animals. Yet even if so – even if all three of the issues mentioned here were resolved in Davis's favor – the Davis diet would still not displace traditional veganism as the dietary embodiment of Regan's philosophy.

More than just animals are accidentally harmed by agriculture and food production. People die as well. And if we are going to count every last field mouse run over by the combine harvester, surely we should also weigh harm to humans in the scale. Leave aside for the moment how many animal deaths one human death is worth. Instead, let us start by simply asking if there are aspects of beef production that make it potentially more hazardous to human beings than crop harvesting.

I can think of four. The first is manure, which is a carrier of *E. coli*, cryptosporidium and other causes of disease. In 2000, seven people in Walkerton, Ontario died from *E. coli* poisoning after manure from a cow farm leaked into the town's improperly filtered water supply (O'Connor 2002). A 1993 outbreak of *Cryptosporidium* infection in Milwaukee, meanwhile, killed 100 people.[9] Grass farmers sometimes say that when their animals are fed grass rather than grain it eliminates the possibility of *E. coli* infection, but the only scientific study on this issue of which I am aware found that there is no significant difference between the amount of the potentially deadly strain of *E. coli* produced by grain- and grass-fed cows (Fegan et al. 2004). Regardless, it seems reasonable to assume that compared to a vegetable crop system – which can be fertilized without manure – Davis's model will result in the greater chance of human harm, due to the higher probability of catching any of a variety of manure-borne diseases.

Then there are farm accidents. An old show business adage says never work with children and animals, because both are unpredictable. The truth

[9] It has never been shown that cattle feces was the cause in this particular instance; I mention it only to give a sense of the damage *Cryptosporidium* can do. For more information, see Wisconsin Department of Natural Resources (2006).

of that remark regarding animals is confirmed by a National Institute for Occupational Safety and Health study which found that farmers that work with beef, hogs and sheep had more lost-time injuries and restricted workdays than farmers who raise cash grain, field crops, fruits, vegetables, nuts or nursery-raised food (NIOSH 1993). A second study, documenting animal-related workplace fatalities, found that U.S. cattle kill approximately 24 people every year. The most common methods of death were either being directly attacked (especially by bulls) or by being crushed by a falling animal. One person died of cardiac arrest after being hurt by a stun gun used to direct cattle, while 15 others died when their cars hit cows that had escaped from a field (Langley and Hunter 2001). We can only imagine what the figure would be if the current number of cattle were not only doubled, as Davis's model encourages, but released from the cramped feedlots that currently typify the U.S. beef industry and allowed to graze in pastures.

Food animals need to be killed, and that requires slaughterhouses. Working in one has been called the most dangerous job in the United States (Schlosser 2002: 172). In large part this is because most of the work must still be done by hand. While chicken slaughter plants are easily mechanized due to the breeding of uniformly sized chickens, cattle come in all sizes and shapes, and vary in weight by hundreds of pounds. As Eric Schlosser notes, "the lack of a standardized steer has hindered the mechanization of beef plants. In one crucial respect meatpacking work has changed little in the past hundred years . . . the most important tool in a modern slaughterhouse is [still] a sharp knife" (2002: 173). As a result, not only do slaughterhouse employees often stab themselves or other workers, but because of the repetitive nature of their work, they have the highest cumulative trauma injury rate in American industry, 33 times the national average (Bureau of Labor Statistics 1999).

Finally, there is methane. When cows belch or pass gas they release it into the atmosphere. According to the Environmental Protection Agency, U.S. cattle alone emit 5.5. million metric tons of methane per year, and cattle account for 19% of the globe's human-related methane emissions.[10] Methane is a greenhouse gas, and there is widespread concern about the

[10] U.S. Environmental Protection Agency (2006a). The FAQ figures are derived from U.S. Environmental Protection Agency (2006b). A separate study found that grazing management practices can reduce methane emissions by 22% (see DeRamus et al. [2003]). This study does not address whether methane-reducing management methods would be feasible on a wide scale, but even if so, it would not change the point at hand, which concerns the greater amount of methane produced by any type of beef farming compared to crop and vegetable cultivation.

contribution such gasses make to global warming. Cattle agriculture on the scale Davis envisions could well prove to be a harm, not only to the environment, but to human beings as well.[11]

What difference do these four possible sources of human harm make? No doubt farmers who grow wheat are sometimes crushed when their harvesters flip over on them. But I take the aforementioned potential sources of harm to show that raising beef is likely to kill at least one more person a year than would an agriculture system not involving cattle. This is important to note given that Davis's goal is to offer an immanent critique of Regan. If Regan ranks the death of even one normal human adult as infinitely more serious than the death of an animal, that would provide Regan with strong grounds for rejecting Davis's model. Even though Davis's model might save a few animal lives, Regan could reply, it is not worth it if it involves more dead people.

Regan does in fact rank the death of a person as far worse than the death of any animal. This is hardly surprising. Few beliefs cut more deeply in our consciousness than the notion that killing a person is worse than killing a mouse, and it is difficult to imagine a theory of animal protection being plausible if it implied that human and animal deaths were morally indistinguishable. Regan demonstrates how his theory respects this commonsense belief in a discussion of a lifeboat that only has four seats. If we have to choose between five potential passengers, four of them ordinary humans and one canine, who should be thrown overboard? Regan gives the same answer any normal person would, namely, the dog. Crucially, however, Regan goes on to note that his answer would be the same even if there were many more dogs. Indeed, it is worth stressing that for Regan, killing one person is worse than killing any number of animals. "Let the number of dogs be as large as one likes; suppose they number a million," Regan writes. "The million dogs should [still] be thrown overboard and the four humans saved" (Regan 2004: 325).

This passage illustrates the principle Regan urges us to take into account when deciding between two options that involve different levels of harm to human beings. If killing an infinite number of animals would prevent one

11 It is interesting to consider (as one anonymous reviewer suggested) whether a purely plant-based agricultural system would require more tractors and trucks than Davis's model, and so produce a greater amount of carbon dioxide. To answer this question we would presumably need to conduct a detailed comparison of the amount of carbon dioxide produced by tractors and harvesters in vegetable and wheat fields versus the amount produced by vehicles travelling to and from slaughterhouses. In addition, we would need to take into consideration the larger amount of plant crops in a vegan system which would convert carbon dioxide into oxygen through photosynthesis. Such factors lead me to suspect that a vegan model would not in fact produce more carbon dioxide.

human death, then we should kill the animals. Regan is happy to make this allowance, for as he notes, the great majority of the harms currently inflicted on animals are not done to save human lives, but simply to provide entertainment (as in hunting) or because people prefer the taste of chicken and other meats over equally nutritious vegan food.[12] As Regan puts it, "What the rights view implies should be done in exceptional cases – and prevention cases, including lifeboat cases are exceptional cases – cannot fairly be generalized to unexceptional cases" (2004: 325). Given these four harms listed, raising beef will likely kill more people than raising vegetables. The vegan model can therefore be justified as a prevention case, and Regan once again wiggles out of Davis's trap.

Of course, Regan's discussion of the lifeboat scenario can be criticized. Peter Singer, for example, has asked whether Regan's endorsement of killing animals to prevent harm to people is compatible with Regan's total ban on animal experiments, some of which are conducted in the hope of curing fatal human diseases (Singer 1985a; Singer 1985b). But even if we grant Singer's criticism, which has the effect of making experiments on animals ethically justifiable under certain circumstances, that is irrelevant to Davis's argument with Regan about food animals (it would only be relevant if we were deciding between eating an animal and eating a person). Moreover, as with the predation exception, offering this criticism would substantially change the nature of Davis's criticism. It would again involve sacrificing the immanent nature of his argument, and replacing it with a direct challenge to one of Regan's premises.

Accidental Versus Deliberate Killing

A defense of traditional veganism that stresses the human harms of beef cultivation may seem strange. The goal of Regan's book, after all, is to offer a theory of animal rights. Isn't it somehow missing the point to mount a defense of *veganism* of all things, that stresses the profound superiority of human life? Doesn't this render irrelevant the killing of animals, not only in wheat and cow fields but also in slaughterhouses?

A human-centered defense of veganism strikes me as no more outlandish than a case for beef eating justified in the name of animal protection. Moreover, this objection overlooks the fact that classifying grass-fed beef as a prevention case treats it as an exception. Even while advancing the

[12] The nutritional value of vegetarian food is documented in American Dietetic Association (2003). The nutritional value of vegan diets is discussed in Mangels and Havala (1994).

human-centered argument against eating grass-fed beef, a proponent of Regan's theory can continue to make direct appeal to the value of animal life when condemning hunting or eating chicken and pork – not to mention most U.S. beef as it is raised now. In short, Regan's primary animal-centered argument can still be made against nearly all types of animal harm.

However, the uneasy feeling that accidental harms to people shouldn't be the decisive factor is nonetheless an intuition worth pausing over. What it should draw our attention to is a key premise of Davis's argument, namely, that there is no moral distinction between accidental and deliberate harm. As he writes in contrasting deliberately killed cows and accidentally killed mice, "the harm done to the animal is the same – dead is dead" (Davis 2003: 392). This notion is central to Davis's criticism of Regan, as it is what justifies calculating accidental and deliberate harms as indistinguishable wrongs.

However, Davis makes a strange remark in justifying his equal ranking of the two forms of killing. It occurs in the following passage: "[Angus] Taylor says about the questions of intent, 'A utilitarian is likely to see no moral difference between the two, since utilitarianism holds that it is consequences that count and not intentions'" (2003: 392). The reference to utilitarianism is strange because Regan's argument is based on deontological rights theory, utilitarianism's great modern rival. Davis does not cite any passage in which Regan himself calls into question the distinction between accidental and deliberate killing, and I am unaware of any instance where Regan does so. So Davis's immanent critique, it turns out, silently depends on a premise that Davis himself introduces. The real question his paper raises, then, is whether it is plausible to say there is a difference between accidental and deliberate harms.

I believe there is. In most legal systems, the difference between accidental and deliberate killing is the difference between manslaughter and murder. Applied to animals, surely we recognize a distinction between accidentally hitting an animal while driving on the highway and intentionally backing over it with the express aim of ending its life. Even a utilitarian can grant greater weight to intentional killing, on the grounds that lethal intentions can have more adverse effects that non-lethal ones. At the very least, it is far from clear that a rights-theorist such as Regan, who urges us to take seriously the moral claims of animals, is obliged to say there is no difference between accidental and deliberate harms to animals. If we take the moral claims of human beings seriously and we draw a distinction between accidental and deliberate killing in their case, why can't an animal advocate say the same thing when it comes to field mice?

This highlights the real significance of unintended human deaths caused by manure and other sources. They are a consistency test for Davis's argument. Davis judges Regan's argument by a standard that leaves no room for ranking accidents as lesser wrongs than intentional killings. Applying the same standard to Davis's argument, we see that it creates problems for his account as well. Only in Davis's case the accidents involve people, not animals. But Davis never gives us grounds for distinguishing between accidents involving people but not those involving animals. Indeed, the quote he offers from Taylor could be read as challenging the significance of the distinction in the case of both animals and humans (which would surely be an extreme view).

Although Regan does not rank animals on a par with people, his theory does urge us to extend many common moral notions we reserve for human beings, such as rights, to other creatures. By that standard, the most plausible version of Regan's theory would be one that does make a distinction between accidental and deliberate deaths, in the case of both people and animals. If so, then the debate between Regan and Davis hinges on whether our everyday habit of distinguishing between deliberate and accidental harm makes sense. The more Davis seeks to preserve his challenge to Regan, the more he will have to attack this common-sense and, to my mind, reasonable distinction. Moreover, Davis will have to attack it in such a way that accidental deaths become just as important as deliberate ones in the case of animals, but not in the case of human beings. In other words, Davis must show that the accidental death of field mice is no different from deliberate killing and so important that it has ramifications for the agricultural sector as a whole; yet at the same time, he must maintain that the accidental deaths of human beings from *E. coli* and other aspects of beef production are not to be judged by such a high standard. Surely that would leave Davis, rather than Regan, with the less plausible argument.

Vegemite and Venison Veganism

Since Davis made the case for burger veganism, numerous variations on his argument have appeared. One such proposal, framed in utilitarian terms, cites Davis's estimates of field animal mortality to defend not only eating cows but free-range sheep and goats as well, while another cites Davis's calculations to defend hunting and killing wild animals (Schedler 2005; cf.

Keller 2005; Cahoon 2009; Demetriou and Fischer 2018). Both proposals presuppose the accuracy of Davis's calculations. My criticisms of Davis's estimates of field animals deaths will therefore have bearing on these cases, and so I leave them to one side (the case for hunting will reappear in my discussion of the doctrine of double effect).

I focus instead on two variations on Davis's idea that offer stronger arguments for an obligation to eat some form of meat. The first variation draws its strength from the fact that while it is similar in sprit to Davis's argument, it cites independent empirical data to make its case. The second cites Davis but does not presuppose Davis's figures are exact. It therefore requires its own separate handling, independent of my discussion of burger veganism.

Michael Archer makes an argument similar to burger veganism that appeals to independent data about field animal deaths (2011a; 2011b). Archer's argument takes as its premise fundamental protectionist claims, including anti-speciesism and equal consideration. Based on data from Australian farms, Archer estimated that at least 100 mice are killed per hectare per year to grow grain there.[13] However, these deaths were not from tractors, but from poisons. Australian farms are periodically overrun by mouse plagues, and farmers use rodenticides that kill about 80% of the mice present to avoid excessive losses to the relevant commodity. As Archer writes:

> Each area of grain production in eastern Australia is subject on average to a mouse plague every four years (Singleton et al. 2005, Caughley et al. 1998). Mouse numbers rise to at least 500–1000/ha or more during these plagues (Singleton et al. 2005). Poisons used to control these plagues kill at least 80% of the mice present (Caughley et al. 1998) (2011a: 980).

Archer employs the formula of (500/4) × 0.8 to arrive at his estimate of 100 mice deaths per hectare of grain. Each hectare on average produces 1.4 tonnes of wheat, and 13% of wheat is edible protein. This results in 55 field animal deaths per 100 kilograms of protein. Australian grazing cattle on average yield 45 kilograms of protein, which means 100 kilos of edible protein requires 2.2 animal deaths. Archer therefore concludes that a diet

[13] Singleton et al. note that while mouse plagues have been reported in China, they occur primarily in Australia, particularly south and eastern Australia, and are rare in Western Australia and Tasmania (2005: 619–20). According to the Australian Government, "The majority of Australian wheat is sold overseas with Western Australia the largest exporting state. The major export markets are in the Asian and Middle East regions … Wheat grown for domestic consumption and feedstock is predominately produced on the east coast" (Australian Government: 2017). This suggests that Archer's analysis pertains primarily to Australians, but nevertheless remains important.

based a concern with animal well-being will not be purely plant based, but also include some free-range beef. Although he does not make precise calculations for lamb and kangaroo, Archer suggests that the same will be true of these meats as well (2011a: 981). Let us term Archer's argument *vegemite veganism*, after its specifically Australian application.

Donald Bruckner offers his own variation on burger veganism. Bruckner cites Davis in arguing that vegans are morally obliged to eat some roadkill. Bruckner takes for granted an argument that he terms the Factory Harm Argument. He outlines its premises and conclusions as follows:

P1: Factory Farming causes extensive harm to animals.

P2: This harm is unnecessary.

C1: The practice of factory farming causes extensive, unnecessary harm to animals.

P3: It is wrong (knowingly) to cause, or support, practices that cause extensive, unnecessary harms to animals.

P4: Purchasing and consuming meat originating on factory farms supports the practice of factory farming.

C2: Purchasing and consuming meat from factory farms is wrong.

The factory harm argument has impeccable protectionist credentials. Insofar as it rules out eating factory-farmed meat, other foods it may permit include humanely slaughtered animals and plants. Bruckner leaves aside humanely slaughtered animals, noting that even humane slaughter systems cause some harm to some of the animals they process. As for plants, he suggests that they are not the only food source that the factory harm argument obliges us to eat. "The premises of the Factory Harm Argument, I claim, support eating roadkill at least as much as they support eating vegetables (2016: 33).

Bruckner specifies that he does not have in mind every animal that is smeared across the highway. His discussion is confined to "large, intact, fresh, and unspoilt animals" (33). He mentions North American species such as deer, moose, and elk, which he collectively refers to as venison, but the argument will also justify consuming other roadkilled animals in different parts of the world. Bruckner suggests 2.1 million is a conservative estimate of the total number of deer, moose, and elk killed on U.S. roads each year (34). To be conservative, Bruckner leaves aside the fact that elk and moose are larger than deer, and assumes that all 2.1 million are deer, and that only 75% of them are found in a condition suitable to be eaten. That would amount to over 41 million pounds of ethically sourced meat.

Bruckner accepts that the criticisms of Davis already noted, several of which I first outlined elsewhere, call into question the argument for burger veganism (Lamey 2017). But even if Davis's figures are off, Bruckner writes, there is no dispute that at least some animals are harmed in plant food cultivation. Road-killed meat is a morally superior alternative:

> Road-killed animals are already dead, so they are not harmed by picking them up as livestock are harmed by common husbandry practices. And no animals are killed in the process of picking up roadkill, as field animals are killed in the process of crop farming. So on the very standards of the Factory Harm Argument, we are obligated to collect and consume roadkill. (2016: 36)

Bruckner improves on Davis by not basing his argument on precise calculations regarding overall field animal mortality. His argument, which we can term *venison veganism*, rather presupposes only that some animals are harmed in vegetable and crop cultivation. Insofar as eating dead animals we find on the side of the road harms so animals whatsoever, protectionist arguments for not eating meat obliged us to replace some of the plants we eat with road-killed meat.

What are we to make of vegemite and venison veganism? To start with the former, it is based on an innocent yet noteworthy mistake. The mistake in question concerns the scope of mouse plagues.[14] Archer writes that "each area of grain production in eastern Australia is subject on average to a mouse plague every four years" (2011a: 980). The sources Archer cites support this statement insofar as "each area" is understood to refer to large geographic regions such as states or parts of states (Victoria, southern Queensland, etc.) (Caughley et al. 1994: 11; Singleton et al. 2005: 618). But Archer's calculations take "each area" to refer to every hectare of grain cropland in Australia. The difference here is between a quadrennial mouse plague happening somewhere in all of Victoria and such a plague happening on every last Victorian farm. This mistake undermines Archer's calculation of mouse poisoning deaths. A more accurate picture is suggested by the Cooperative Research Centre for Pest Animal Control, which notes that each year between 100,000 and 500,000 hectares of grain crops in Australia are subject to mouse plagues (McLeod 2004: 9). Australia plants

[14] I owe this point to Animal Liberation South Australia (2014c). For additional empirical criticisms of Archer by Australian animal rights advocates, see Animal Liberation South Australia (2014a), (2014b), and All Animals Australia (2014).

approximately 22 million hectares of grain crops annually (Australian Export Grains Innovation Centre 2016). These figures suggest that in an average year 2.3% of Australian grain cropland is hit by plague. When Archer's figure of 55 deaths per hectare of grain is recalculated to only apply to 2.3% of crop land, the mortality rate for grain becomes 1.27 animal per hectare.

What is more, mouse plagues are not confined to crop fields. Although the plagues are particular to Australia's grain belt, the mice's spawning grounds extend beyond individual wheat fields. That is why models for predicting mice plagues are based on the food available to mice not only in fields but also in grazed pasture (Pech et al. 1999: 81). As the Government of New South Wales explains, "in the field, mice are always present, but mostly in low numbers. Refuge areas such as channel banks and the more densely vegetated pastures are ideal habitats, where detection is difficult" (2017: 3). If mice are not confined to crops when they spawn, the same is true when it comes to overrunning farms. Upon reaching plague proportions they attack agricultural operations indiscriminately, as the Cooperative Research Centre report describes:

> Mice invade intensive farming enterprises such as poultry housing and piggeries causing damage to infrastructure, spoiling feed and, in some cases, causing damage to animals (Caughley et al. 1998). In addition to intensive farming, the grazing industry is impacted through mice consuming pastures, destroying feed grain and damaging stored hay. Similarly to grain producers, farm equipment and buildings are damaged as a result of a plague. (McLeod 2004: 11)

The point to note here is that mouse plagues also damage grazing land. A primary reason farmers resort to poison is to avoid the extensive economic losses mouse plagues cause. Insofar as mouse plagues damage farms indiscriminately, it is misleading to argue that farmers poison mice "to grow grain" (Archer 2011a: 980). The poisoning that occurs in Australia is done to protect all aspects of affected farming operations, including pastures. Archer's analysis errs in failing to associate mouse poisoning with animal agriculture. This undermines his attempt to use mouse poison deaths as grounds to prefer pasture-raised meat over plant foods.

This still leaves Bruckner's cheeky and original argument for venison veganism. Bruckner sees his project as fundamentally negative. "My thesis is not that we are obligated to collect and consume roadkill. My thesis is that the usual arguments for vegetarianism imply that we are obligated to collect and consume roadkill" (2016: 43). Bruckner's negative understanding of his project explains why his paper is called "Strict Vegetarianism is Immoral." Bruckner's approach to classifying diets, however, is based on the list-of-ingredients approach. As I have argued, it makes more sense to classify vegan diets according to a common rationale, that of reducing harms to animals. On that basis Bruckner's proposal is itself a form of veganism.

There is an additional reason to see Bruckner's dietary ethic as a form of veganism. Even if the argument were successful, it would justify consuming only small amounts of meat. The U.S. population in 2017 was 325 million. Bruckner estimates that his scheme would permit eating 41 million pounds of venison, which would supply each American with 0.12 pounds, or less than two ounces, of meat per year. Even if Bruckner's estimate were off by a factor of ten, that would still justify an average consumption of only 1.2 pounds of venison per year. In practice, most people would eat no meat in a given year. This is because Bruckner's proposal would condone drivers and car passengers eating an entire deer that they salvaged from the road. Bruckner estimates that each deer yield on average 35 pounds of meat. So every time one person brought home and consumed an entire deer, 29 other people would be denied their annual portion of venison (assuming 1.2 pounds per year). Alternatively, Bruckner's scheme might be implemented in a way that permitted only consuming our fair share of roadkill, with the rest redistributed to others. Either way, even if the argument succeeds it entails consuming only small amounts of venison and so represents a minor change to veganism rather than a fundamental challenge.

But does the argument succeed? Four considerations suggest it does not. The first is highlighted by Julia Driver, who asks whether an ethical vegan could reasonably eat meat other people had already thrown out. We might think so on the familiar grounds that such a habit would have no impact on the fate of animals. Driver however suggests that if meat eating is wrong in general, then it is also wrong for the vegan dumpster diver. "A system is established whereby some people habitually benefit from the misdeeds of others, allowing them to reap the benefits without the dirty hands" (Driver 2016: 76). Extended to roadside animals, if some were killed by drivers guilty of negligence (or worse) then the same problem of "bypassing norms" will also apply to consumers of roadkill: they will benefit from, and so be complicit in, an immoral practice (76).

The case for venison veganism is also undermined by a rule-following problem identified by Neil Levy. Levy defends strict dietary codes by citing psychological research on how we apply behavioral rules. The research suggests that "the extent to which rules must be interpreted in order to be applied predicts their liability to encourage self-deception" (2016: 180).[15] Telling ourselves that we are only going to eat free-range meat, for example, requires us to make ongoing judgement calls. To take one of Levy's examples, "should I accept the waiter's assurance that the pig was slaughtered humanely?" (180). The science suggests we have a tendency to succumb to temptation in answering such questions. A strict no-meat rule by contrast does a better job managing weakness of will and requires less cognitive labor. Whether roadkill was served by waiters or sold out of the back of trucks, people who acquired it would inevitably have to decide whether or not the meat they wanted to eat really was roadkill, risking a demand for animals deliberately killed for food.

Levy's defense of inflexible dietary codes finds some support in a study comparing how well vegetarians and conscientious omnivores did in living up to their respective dietary codes (Rothgerber 2015). The conscientious omnivores sought to eat meat that satisfied certain ethical standards, e.g., it came from animals that were raised and slaughtered humanely. The study found that they were less attached to their code than vegetarians were to theirs. Conscientious omnivores "reported violating their diet more, felt less guilty when doing so, believed less in animals rights, were less disgusted by factory farmed meat [and] rated its sensory characteristics more favourably" (ibid., 196). Bruckner's argument presupposes a strong commitment to animal rights, and so his dietary rationale is unlike that of the conscientious omnivores who participated in the study. The study's central point however was that it is more difficult to follow conscientious omnivorism, and it is reasonable to wonder whether venison veganism might face the same problem.

Consider just the fact that conscientious omnivores have a more positive reaction to the sensory characteristics of meat. One explanation, admittedly speculative, is suggested by reports from former meat eaters who turn vegan that their taste buds change and as a consequence they no longer desire meat. This raises the possibility that someone who swore off all meat save roadkill would still be "weaponizing" their taste buds, insofar as they would continue to cultivate the desire for meat that the study suggests makes conscientious omnivorism a more challenging diet to implement than vegetarianism or veganism.

Brucker might mitigate some of these problems by only requiring us to eat animals we run over ourselves. This however raises a third problem. His

[15] See Ainslie (2001).

rationale for eating roadkill is that it will cause fewer animal deaths than agriculture. Even if the number of harvester deaths turn out to be small, he argues, it will still be more than the number of deaths caused by eating roadkill, which must be zero. But would a norm of eating accidentally killed animals really not cause any additional deaths?

Grounds for doubt came to mind when my wife and I had to decide whether to drive or fly on a family trip. We weighed the usual pros and cons of money, time, and screaming kids in the back seat. Bruckner's scheme would add a consideration in favor of driving: we might happen to come across some dead deer along the way. His proposal thus incentivizes people to take to the roads in hope of killing animals, thereby gaining access to the rare delicacy of their flesh. This suggests that we cannot assume zero additional animal deaths after all. We rather have two food-production schemes, agricultural and agriculture plus roadkill, both causing an unknown number of animal fatalities. This undermines Bruckner's central claim that venison veganism will necessarily harm fewer animals that the traditional version.

Venison veganism faces one final problem. Bruckner's claim that the case for it presupposes "the usual arguments for vegetarianism" is not quite right. A commitment to anti-speciesism is widespread among protectionist authors. This means that on a protectionist view, the question of eating animal roadkill will be inseparable from the question of eating human roadkill. Imagine that there were some sinister highway on which it was common for merely sentient human beings to be run over. Assume that they had no relatives who cared about them. Surely few people would see this as an opportunity to eat them. Whatever reasons there are not to eat roadkilled humans with mental abilities no more complex than merely sentient animals, the same reasons will also rule out not eating merely sentient nonhumans. Such a ban might be justified on consequentialist grounds: it would eliminate the aforementioned incentive to seek out roadkill. Alternatively, such a ban might be based on the belief that the opportunistic consumption of a once-sentient being is incompatible with symbolic respect for such a being's dignity (Anderson 2004: 282). My point is not to defend any particular basis for such a ban. It is rather to suggest that some such basis is available. Whatever basis applies in the case of merely sentient humans will on anti-speciesist grounds extend to merely sentient animals as well.

Future Directions of Humane Agriculture

The reaction of many protectionists to burger veganism and its cousins will not be to conclude that we should eat meat in the name of animal rights.

It will be rather that we should take steps to reduce the number of animals harmed by plant agriculture. In this final section, I want to provide some cause for optimism that this can indeed be accomplished.

It is instructive to take note of recent changes that have occurred regarding how some pest species are managed. In Europe, for example, international treaties have promoted wildlife conservation by restricting the hunting of large grazing birds (Nilsson et al. 2016: 164). Where killing was once the default option to limit agricultural damage caused by geese, cranes and swans, contemporary methods now involve non-lethal practices such as scaring campaigns, diversionary fields and refuge sites (Jensen et al. 2008; Nilsson et al. 2016). Such methods were unheard of prior to the 1990s. Although these particular methods may not be directly transferrable to the management of rodents and other small mammals, they illustrate the cultural nature of agricultural pest management. Rather than a brute given, steps taken to manage crop damage evolve in response to changing technology and values. To date conservation has been an influential driver of change where animal rights has not. Nevertheless, there are existing agricultural trends that do promise greater well-being of field animals.

Consider, for instance, alternative tillage practices. No-till methods are popular in South America, where they are used on 50 per cent of cropland in Brazil and over 80 per cent of cropland in Argentina, Paraguay and Uruguay (Gianessi 2014). Currently, no-till methods are used on 34 per cent of cropland in the U.S. (USDA 2005). In Europe, conservation tillage is employed on 25.8 percent of cropland (Kertész and Madarász 2014: 91). No-till and conservation tillage seek in varying degree to prevent soil erosion by minimizing soil disruption. As has been noted, there is debate over the relative impact of conventional tillage on field animals compared to no-till. Still, insofar as no-tillage and conservation tillage reduce the number of times plows and row cultivators pass through crops, they may pose fewer threats to field animals. Although the spread of both practices has been due to environmental considerations rather than any concern with animal protection, they nevertheless represent a well-established trend away from conventional tillage. Informed debate over the deaths of field animals require further research on the possibly varying effects of different tillage practices.[16]

Another existing trend likely to favor field animals concerns indoor agriculture. Greenhouses, for example, are popular in the Netherlands,

[16] This is not to suggest that there are no other costs to no-till and conservation tillage: they do tend to use more herbicides as currently practiced. For discussion see Fischer and Lamey (2018).

where the food grown in them constitutes 22 per cent of the country's total agricultural value (Marcelis and Hemming 2013). Recent years have also seen the birth of so-called vertical farms, which see plant crops grow in indoor structures ranging from skyscrapers to stacked shipping containers. Rationales for the practice are commonly environmental. "We conserve our resources, we recycle our water and nutrients, we don't use any tractors and aren't emitting any hydrocarbons or greenhouse gas as we plant and harvest," states the founder of one U.S. vertical farm (Strasser 2014). Similarly, the managers of the world's largest indoor farm, based in Japan, claim that their operation is 100 times more water efficient than outdoor fields (MacDonald 2015). Be that as it may, indoor farming promises to have an extremely low harm footprint, as there are no animals involved in those systems.

Another existing practice likely to reduce field animal deaths is the use of contraception as a form of pest control. In 2013 New York City began testing a new product to control the rat population in its subways. Bait eaten by female rats made them permanently infertile, to the point that rat populations near subways decreased 43% (Flegenheimer 2013; Molteni 2017). Prior to the subway trial, the rat contraceptive was researched in agricultural locations in Laos, India, and the Philippines, and the company that created the contraceptive is now working on versions targeting mice and pigs. This raises the possibility that at some point in the future non-lethal agricultural pest control may be feasible on a large scale.

There are, in addition, entirely new agricultural practices that may ultimately take off. For instance, there is some evidence that the height of crops can influence the population trends of small mammals that live in fields (Jacob and Hempel 2003: 49). This finding might inform the management of cover or other crops so as to reduce anthropogenic field-animal mortality. Similarly, existing research suggests that the manipulation of adjacent non-crop habitats, such as ensuring vegetation does not grow above 10 centimeters, can naturally reduce the population of field rodents, an approach that might be extended to the management of inter-row plants and weeds in crop fields (White et al. 1998). Fencing practices and agricultural machinery design might also be modified so as to protect field animals. The embryonic field of floating farms, which currently are at the proposal stage and which involve very large vertical farms that float on pontoons, might be encouraged for the same reasons (Smart Floating Farms 2017).

Alternative tillage practices, indoor farming, and rodent contraceptives are existing agricultural practices that have the potential to reduce field animal deaths, and there are others that we might eventually develop. However, none of these practices have received any attention in the conversation about field animal mortalities. George Schedler, for example, writes that "there is no reason to believe a method of commercially harvesting vegetables that causes no suffering to field animals will ever be found. There is no effort underway to discover such a method" (2005: 505). Schedler's remark does not describe current reality. Agriculture has taken a wide variety of forms throughout history, and current trends would seem to raise the serious possibility that plant agriculture might someday kill very few animals – perhaps even none.

Conclusion: After Burger Veganism

In 2016 the last sugar mill closed in Hawaii (Solomon 2016; Kai-Hwa Wang 2016). Recent years had seen a substantial decline in Hawaiian sugar cane production due to increased labor costs and global competition. When the last mill closed, sugar cane stopped being harvested on an industrial scale in Hawaii. When the last mill closed, a uniquely Hawaiian tradition ended with it. Globally, 60% of sugar cane is harvested by hand. Those regions that employ mechanical cultivation most commonly use combine harvesters (Elvers 2017: 968). Hawaiian plantations used a distinctive local method, one that employed v-cutter harvesters and push rakes. Hawaii was thus unique in the particular machinery that it used to cultivate sugarcane.

The historical particulars of Hawaiian sugarcane production are worth recalling when considering the legacy of burger veganism and its dietary offspring. On the one hand the unique method that Hawaiian plantations employed speaks to the difficulty already noted of making claims about field animal deaths that generalize from one crop or species, or in this case region, to another. Given that Hawaiian sugar cane cultivation was historically practiced in a unique form that no longer exists, it is unclear what relevance the deaths of field animals under such a method have to determining field animal mortality rates for other forms of sugarcane production, let alone agriculture as a whole.

But the history of sugar in Hawaii also illustrates the historically contingent nature of agricultural practices. Different methods come and go. When it comes to ethical issues involving human beings, it has long been axiomatic to propose reforming industries and practices that do harm, hence historic campaigns to limit the number of hours laborers can be asked to work, or to limit the economic exploitation of children. It will seem obvious to many protectionists that, rather than reconcile our dietary ethic with existing agricultural practices, the real upshot of burger veganism is that we should reform agriculture in the name of our dietary ethics.

Animal protectionists however can still learn something from burger veganism. As a professor of animal science, Steven Davis entered the protection debate with an intimate familiarity of animal husbandry. One of the challenges of advancing protection theory is that doing so amounts to asking people to accept a lifetime homework assignment regarding their diet. Rightly or not, this can cause people to reject the philosophy. Davis's work, however, raises the possibility of an intermediate step, one that falls short of veganism or even vegetarianism, but which still represents a meaningful dietary change. This occurs when Davis notes that the average slaughtered cow represents 1,200 lbs. of meat, whereas the average chicken only provides 6 lbs. (Hence the eight billion chickens killed every year, compared to only 37 million cattle).

This simple difference between animal carcasses has potentially major implications for efforts to act on protectionist philosophy. On the one hand, it should cause protectionists to recognize that not all meats are created equal. In terms of the number of animals killed, one serving of chicken represents 200 servings of beef. Causing someone to give up chicken (or fish) before beef would thus be a victory in itself. Although eliminating chicken alone should not replace veganism as animal protection's ideal, it deserves serious consideration as the first step people are asked to take when it comes to changing their diet. This is especially so when one considers the large number of people who cut out red meat but keep eating chicken. They may actually be doing more harm than good when it comes to the overall number of animals killed, depending on whether their chicken consumption increases. In short, the issue of carcass size deserves more attention than animal advocates have given it.

It is practical and immediate questions of this kind, rather than large-scale blueprints for national agricultural reform, that represent burger veganism's real contribution. Until such questions receive further consideration, however, we should continue to affirm a traditional conception of animal ethics. A conception that says if we are committed to protecting animals, it follows that we will not eat them.

CHAPTER 4

The Dinner of Double Effect

Introduction

A trolley is hurtling down the track. Tied to the track are five people who will die if you do nothing. You find yourself beside a lever, which, if pulled, will divert the trolley down a sidetrack to which a single person is tied. Your choice is between doing nothing, which will result in five deaths, and redirecting the trolley down the sidetrack and thereby killing one person. Call this scenario *Lever*. Whatever choice you make must be consistent with your choice in a second version of the dilemma, which we can call *Wrestler*. A trolley is once again hurtling toward five helpless people. Only now you are standing on a bridge over the tracks behind a large wrestler. If you push him off the bridge it will save the five innocents, but will again see your action cause the death of another innocent person.

The trolley problem is a familiar staple of introductory ethics classes. The problem is a popular teaching device due to the way it leads careful respondents to consider the issue of greater versus lesser consequences – five deaths or one? – alongside the separate issue of what morally relevant difference, if any, separates pulling the level from pushing the big man. Anyone who has taught the trolley problem knows it is common for a large majority of the class – routinely over 90 per cent in my experience – to recommend pulling the lever in the first scenario. In the second scenario it is more common for a smaller majority to opt *not* to push the wrestler off the bridge. This pair of responses is inconsistent if consequences are all that matter. After all, in both dilemmas, the choice involves one death or five.

The doctrine of double effect (DDE) can render these two responses consistent. In the traditional formulation of the doctrine, it holds that a higher standard of justification is required for actions that intend harm rather than merely foresee it. In Lever, the goal of saving the five lives would be achieved if the person were not present on the sidetrack. His or her death is not essential to the goal of saving the five. In Wrestler, bodily harm to the wrestler is the means by which the five are saved. This difference, proponents of the doctrine hold, means that pulling the lever is permissible while pushing the big man is not.

As we saw in the previous chapter, the animal ethics literature now abounds with variations on the argument for burger veganism. These proposals commonly turn on the same issue Steven Davis highlighted, the death of animals in plant agriculture. The empirical details of the different proposals vary, but a common feature of all versions of burger veganism is that they rank dietary options according to the overall number of animal deaths each diet requires. If eating a diet with free-range grass-fed beef, mutton or kangaroo involves fewer overall animal deaths than a traditional plant-based vegan diet, then that is taken to demonstrate the immorality of traditional veganism.

A problem for this family of views is their failure to note the ethical relevance of the doctrine of double effect. If the DDE is applicable to agricultural ethics then it raises the possibility that it remains immoral to follow a diet based on the deliberate killing of animals, even if doing so did result in fewer animal deaths than a plant-based diet. The reasoning here is similar to that which says it is impermissible to kill the wrestler, even if five lives really will be saved. In both cases, it is because the sheer amount of harm is not the only ethically relevant consideration. The intention behind the killing also matters. It is therefore worth noting the applicability of the doctrine of double effect to the contemporary debate about what type of diet animal protection entails.[1]

In what follows I seek to outline the relevancy of the DDE to the debate over the deaths of animals in crop cultivation. A major challenge of bringing the DDE to bear on any debate involving animals is that the most sophisticated contemporary versions of the DDE employ a rationale that makes the DDE inapplicable to entities that are not persons, a category that plausibly includes most if not all animals. Contemporary versions of the DDE have adopted such a rationale in order to avoid the so-called closeness problem. This problem concerns the degree to which intended outcomes can genuinely be distinguished from merely foreseen

[1] I was made aware of the possible relevance of the doctrine of double effect to the debate by the title of philosopher Daniel Crow's unpublished paper "Dead is not Dead: The Doctrine of Double Effect and Burger Veganism." Although I have not read the paper, I owe my interest in this topic to Crow's title. For previous applications of the doctrine to agricultural ethics see the brief discussion in Bernstein (2004: 31–2) and Cochrane (2012: 96–8). Bernstein and Cochrane both employ a traditional, and therefore inadequate, version of the DDE.

ones. Contemporary proponents of the DDE have avoided the closeness problem by revising the doctrine so that it depends on distinctions other than that of merely foreseen vs. intended harm.[2] I take over and endorse one such new distinction, that of harmful direct agency vs. harmful indirect agency. I then go on to outline a new version of the DDE that avoids the closeness problem but, unlike the best existing versions, is applicable to our dealings with merely sentient animals. I note how justifications for the best existing version of the DDE can be extended to justifying the animal-friendly version of the DDE I defend. I then apply the animal-friendly DDE to harms done to animals in agriculture, and show how it blunts the force of recent arguments for burger veganism. The strongest such argument makes the case for a vegan form of hunting. Although it is not refuted by the animal-friendly DDE my version of the DDE does drain the argument of much of its force, and it ultimately runs afoul of protection theory more broadly. I also note the philosophical support the expanded DDE lends to a ban on infanticide. I conclude my discussion by briefly mooting (rather than arguing for) the possibility that the animal-friendly DDE may be compatible with more than one moral philosophy.

The Closeness Problem

The doctrine of double effect has historically been invoked to justify a particular kind of harmful action. This is an action that causes harm as a side effect of trying to achieve a good outcome. The doctrine has long been used in a military context, for example, to distinguish cases such as the following:

> **Strategic Bomber**
> A pilot bombs an enemy factory in order to destroy its productive capacity. But in doing this he foresees that he will kill innocent civilians who live nearby.
>
> **Terror Bomber**
> A pilot deliberately kills (the same number of) innocent civilians in order to demoralize the enemy. (Quinn 1989: 336)

In both Strategic Bomber and Terror Bomber, a foreseeable harm is done to civilians. The aim of the Strategic Bomber however is not to deliberately

[2] In addition to the versions of the DDE defended by Quinn and Nelkin and Rickless, discussed later, see also Fitzpatrick 2006. For an extensive discussion of contemporary versions of the DDE, see Nelkin and Rickless 2015.

kill non-combatants: he could achieve his goal of destroying the factory just as easily if there were no civilians present. The terror bomber by contrast does deliberately seek civilian deaths. Proponents of the DDE have traditionally distinguished foreseen outcomes from intended ones in explaining why Strategic Bomber is permissible but Terror Bomber is not. As Michael Walzer puts it, "the intention of the [Strategic Bomber] is good, that is, he aims only at the acceptable effect; the evil effect is not one of her ends, nor a means to her ends (Walzer 2006: 153).

Traditional formulations of the DDE have long been faced with a difficulty. Cases of intended harm can all too easily be re-conceptualized as *un*intended. What is to say for example that the Terror Bomber did not in fact aim at killing civilians? Nothing in his action itself rules out it being animated by a different motive, namely that of making it *appear* that the civilians were dead, in order to demoralize the enemy. If his victims came back to life somehow after the war, his goal would still be achieved. In this way, their deaths are reclassified as side-effects of his genuinely intended end. The harm that comes to the wrestler when he is pushed off the bridge can also be reclassified as a side-effect. On this view, the agent's goal is simply to stop the trolley. If the wrestler were to miraculous be uninjured by its impact, the agent would achieve his goal, in relation to which any harm to the wrestler was merely incidental – and so not intended after all.

Of course it would be miraculous indeed for someone to not be hurt by a runaway train. But the point at hand is conceptual rather than empirical. Insofar as it is possible even to conceive the wrestler stopping the trolley without being harmed, this is enough to show that harm to the wrestler is neither the goal of the agent nor a means to his goal. But if Terror Bomber and Wrestler can be reclassified as side-effect cases, then the central purpose of the DDE, that of distinguishing different harm cases, is defeated.

A possible response to the closeness problem is to resist the reconceptualization that gives rise to it. One version of this response is based on the notion of event identity. It maintains that the death of the civilians in Terror Bomber cannot in fact be conceptually separated from the goal of making them appear dead by dropping a bomb on them. There is no difference between the bomb impacting them and their deaths. Rather, references to the bomb's impact and their deaths are two different ways of referring to the same thing. Similarly, in Wrestler, harm to the big man and being hit by the trolley are not discrete events but one and the same (Bennett 1995: 205).

This way of avoiding the closeness problem however faces its own difficulty. The difficulty is that an event can be individuated according to a) who or what is the event's subject, b) what the subject's properties are and c) when then event occurs. So in the case of Terror Bomber, a civilian might be struck by bomb shrapnel and die an instant later. The civilian subject has different properties, being alive and then dead, at two slightly different times. That is enough to differentiate the bomb's impact from the person's death. Similarly, the slight delay in the wrestler's death after being struck by the trolley can similarly be used to differentiate death from impact.

To avoid this limitation of the Event Identity response, one might shift to a notion of Act Identity. It holds that the Terror Bomber's action is numerically identical to killing the civilians. In other words, killing civilians is simply what the Terror Bomber does. Here the identity is grounded in the fact that it is impossible to intend the bombing without intending the killing. By the same standard, it is impossible to intend the wrestler be hit by the trolley without intending that he be harmed. The problem with the Act Identity response however is that it does not afford ground to distinguish Terror Bomber from Strategic Bomber or Wrestler from Lever. This is because the Act Identity response does not admit of a difference between intending an outcome and merely foreseeing it. This again defeats the purpose of the DDE.

Two Forms of Harmful Agency

A more promising response to the closeness problem revises the DDE. The revision distinguishes between two forms of harmful agency, direct versus indirect.

Harmful Direct Agency: Agency in which harm comes to some victims, at least in part, from the agent's deliberately involving them in something in order to further his purpose precisely by way of their being so involved.

Harmful Indirect Agency: Agency in which harm comes to some victims, but in which either nothing in that way is intended for the victims or what is so intended does not contribute to their harm. (Quinn 1989)

The distinction between harmful direct and indirect agency is similar to the distinction between intended and merely foreseen harm, insofar as the two forms of harm are distinguished on the basis of the agent's intentions. Crucially, however, the new distinction hinges on whether the agent *involved* the victim as a means of achieving her end. Involvement is a more general notion than harm. Thus it does not matter whether the Terror Bomber really intended only to demoralize the other side, or whether the goal of pushing the wrestler off the bridge was to stop the trolley rather than kill him. In both cases the harm was the result of a decision to involve the civilians or wrestler as a means of achieving an end. In Lever and Strategic Bombing, by contrast, the harm that comes to the person on the side track and to the civilians is not the means by which the agent in either case brings about her goal. In Lever the five people on the main track would live even if the single person was not on the side track. Similarly, the death of civilians does not advance Strategic Bomber's end. In both these cases, the death really is a side effect of the goal at hand.

The Revised DDE

The direct-indirect harm distinction has been invoked to reformulate new versions of the DDE, one by Quinn (1989) and one by Nelkin and Rickless (2014). All three writers invoke a Kantian rationale for their revised doctrines. As this particular rationale is a central subject of concern, it will be helpful to employ a label that refers to Quinn, Nelkin and Rickless in particular, rather than any broader group of Kantians or deontologists. Let us call them *California Deontologists* (because all three writers are or were based in California). Although there are significant differences between their positions, for present purposes these are less important than their similarities, particularly as they concern justifying any version of the DDE.

Quinn's version is put forward as an improvement on the traditional version while Nelkin and Rickless's version improves on Quinn. Here we can skip over Quinn's iteration and focus on that of Nelkin and Rickless, who call their revised version DDE-R. To my knowledge it is the most cogent version of the DDE on record.

DDE-R: In cases in which harm must come to some in order to achieve a good (and is the least costly of possible harms necessary), the agent foresees the harm, and all other things are equal, a stronger case is needed to justify harmful direct agency than to justify equally harmful indirect agency (Nelkin and Rickless 2014: 137).

DDE-R innovates by building into the principle several features not found in previous versions. One of these is the reference to possible harms that are least costly. To see the significance of this element, imagine that we actually have three choices in Wrestler. In addition to the options of doing nothing and pushing the wrestler off the bridge, there is also a boulder that we can nudge that will fall on the track. The boulder will stop the trolley and no one will be killed. Of the three options now available, nudging the boulder is the least harmful. If we push the wrestler when the boulder option is available, our action will be wrong: we could have saved the lives of the five people on the track in a less costly way. Were someone to point this out to us, appealing to any version of the DDE would do nothing to justify pushing the wrestler. The description of the DDE-R quoted explicitly acknowledges this limitation. Its applicability is confined to cases such as the original trolley scenarios, where the harm we are considering causing is the least costly of possible harms that we might inflict.

The DDE-R also contains a reference to other things being equal. To understand this aspect consider a revised versions of the original pair of trolley problems. In these new versions there are 50 people on the main track in both Lever and Wrestler. However, in Lever, rather than one person being tied to the side track, there are 49. The outcome of the two scenarios are thus as follows: in Lever, by pulling the lever only one life will be saved. In Wrestler, pushing the big man will save 49 lives. Surely the numbers matter to our overall assessment of which actions are permissible. In Lever, it seems plausible that pulling the lever would remain justified. But the judgment in Wrestler might well change. Whereas when there were only five people on the track pushing the wrestler was impermissible, it seems more plausible to say that killing him is permissible to save 49 lives.

The revised trolley scenarios are a reminder that endorsing a non-absolutist version of the DDE such as DDE-R is not the same as saying that harmful direct agency is always unjustified. Whether it is at the point of saving 49 or some still higher number of lives, there will be a point at which direct harm to the wrestler is justified. Similarly, one can imagine forms of harmful indirect agency that are so costly that they are unjustified. The R-DDE is not meant to deny these considerations. Again, it is meant to apply to cases such as the original pair of trolley problems. In such cases, the harm done to the person on the sidetrack and to the wrestler is equal. It is in cases of equal harm such as these that the DDE-R says there is a higher standard of justification for direct harm.

DDE-R is an attractive moral principle. A primary reason it is attractive is due to a feature it takes over from Quinn. It avoids appeal to the traditional distinction between intended and foreseen harm and instead relies on a distinction between harmful direct versus indirect agency. In doing so DDE-R avoids the closeness problem. The appeal to direct and indirect harm obviates the need to engage in hair-splitting debates about how to conceptualize an agent's precise intention. Given the exhausting debate over the closeness problem, this is no small advantage.

DDE-R, however, has a less attractive feature. This concerns the rationale California Deontologists have offered for it. It is one that would appear to exclude merely sentient beings from the protection of the DDE. As Quinn puts it:

> The doctrine reflects a Kantian ideal of human community and interaction. Each person is to be treated, so far as possible, as existing only for purposes he can share. This ideal is given one natural expression in the language of rights. People have a strong prima facie right not to be sacrificed in strategic roles over which they have no say. (1989: 350)

Setting the DDE's boundary of applicability at "each person" would exclude all animals that are not persons. Setting it at the borders of the "human community" will exclude all animals, period. Either way, the DDE would have no bearing on our moral interactions with the great majority of living creatures.

The rationale that Nelkin and Rickless offer would seem equally unpromising in terms of justifying a moral principle with bearing on our interactions with animals. "On the strongest version of the Kantian rationale, the presumption against harmful direct agency derives from the fact that such agency infringes an independent right not to be made to serve others' purposes without one's consent" (2014: 133). Given that animals cannot consent to our dealings with them, this articulation of the Kantian rationale for the DDE would again exclude animals from its purview.[3]

Traditional versions of the DDE are not as exclusionary as the person-centered versions endorsed by California Deontologists. Older, Catholic versions of the DDE extended to non-persons. Much of the modern philosophical interest in the DDE is due to Philippa Foot's article "The Problem of Abortion and the Doctrine of Double Effect" (1967). Foot's discussion is prompted by the use of Catholic thinkers to distinguish between two forms of harm to a fetus:

Hysterectomy: A pregnant woman has cancer of the uterus. Her uterus must be removed if she is to live.

[3] An alternative possibility might be to say that Nelkin and Rickless's version of the DDE does apply to animals because what matters is that the animals did not consent. This however suggests that animals have an "independent right not to be made to serve others' purposes," which seems far too strong: it is effectively a version of the No-expense Principle, which I discuss later. Such a principle would rule out taking a dog for a walk or riding a horse, even when these activities are done without harming the animal involved.

Craniotomy: A pregnant woman will die unless the head of the fetus she is carrying is crushed.

Foot observes that on a traditional Catholic view, performing the hysterectomy is permissible but administering the craniotomy is not (1967: 6). These cases continue to be widely used examples in contemporary discussions of the DDE, even by proponents of DDE-R and other versions that do not apply to fetuses.[4] As Quinn notes, the abortion cases continue to be discussed by philosophers who reject the Catholic view of abortion is that they naturally illustrate the closeness problem (1989: 336).

The version of the DDE I will shortly defend will differ from both the Kantian and Catholic versions in its range of application. My version will be applicable to a wider category than persons, that of sentient beings. But it will not apply to fetuses as such. An implication of my view is that when fetuses reach the point at which they become sentient they will have moral standing. Prior to that point in their development however they have no such standing on a sentience-based account. Thus being a fetus is not a sufficient condition of being protected by my version of the DDE. It may perhaps also be worth mentioning that if some fetuses have moral standing, it does not necessarily follow that aborting them is impermissible. A wide range of views note the possibility of abortion being permissible even when sentience is taken as a condition of moral standing (Jenni 1994; McMahan 2002; Nobis 2016).[5] My view, while it therefore has some small bearing on

[4] Quinn, and Nelkin and Rickless, gesture in passing toward the possibility of the Kantian versions of the DDE extending to non-persons. Quinn at one point moots the possibility of fetuses being able to consent to harm. This occurs in his discussion of the Kantian rationale and its strong presumption that voluntary consent is a prerequisite of using a person to advance a goal of one's own. The "chief exceptions" to this presumption are cases in which an individual has a moral responsibility to give themselves up to the goal of others. He mentions Craniotomy possibly being such a case. "I suppose it is at least open to question whether the fetus in [Craniotomy], if it could grasp its predicament, would have to accept, for the sake of its mother, the sacrifice of its life" (Quinn 1989: 349). This passage is puzzling. A fetus that could not only grasp what craniotomy is but consent to it would be exercising a defining feature of personhood. If it could do so then surely it would be entitled to the protection of the DDE in its Kantian forms. However, it is biologically beyond question that fetuses cannot do this, so the purpose of Quinn's counterfactual is unclear. For their part, Nelkin and Rickless note that the significance of the abortion cases depends on background assumptions about the moral standing of fetuses. The idea that there is a morally relevant difference between the abortion cases will find support "only among those who think that human fetuses are persons (or near persons)" (2015: 406 fn. 9). The reference to near-persons might be though to open the possibility of extending some version of the Kantian DDE to fetuses. On my reading of the California Deontologists, however, there is no basis on which to extend either of their versions of the DDE to entities that are not persons. Not only fetuses and animals, but newborn infants and human beings with congenital mental handicaps that render them merely sentient would all seem excluded from any version of the DDE based on the Kantian rationale.

[5] Lee et al. (2005) suggest that fetuses become sentient some time near the end of the second or beginning of the third trimester. The Centers for Disease Control and Prevention states: "In 2012,

the abortion debate, does not entail a restrictive view regarding the moral permissibility or legality of abortion.

The version of the DDE I propose is supplementary rather than revisionary. I do not seek to challenge the applicability of the DDE-R to cases involving persons. Rather I want to suggest that there is an additional true principle that is applicable to merely sentient creatures. Nelkin and Rickless anticipate such a possibility: "It may be that there is an even more general principle of which DDE-R is just one more specific version" (2014: 149). My view is consistent with there being a more general principle of which DDE-R and the animal-friendly DDE are both more specific versions. Defending a supplementary version of the DDE however is philosophical challenge enough.

Animal Purposes

I propose a new rationale for a supplementary version of the DDE.[6] On this rationale the DDE applies to our dealings with sentient beings that can form purposes. The distinction between persons and sentient beings remains morally relevant, but being a person is not a prerequisite of being a subject to whom the DDE can apply.

I take it for granted that animals can pursue purposes of their own. To say this of course is not to say that they can pursue the same purposes as persons. The precise nature of the abilities of diverse animal species is an open empirical question. It does not seem open to question however that many animals engage in goal-directed behavior, as when they escape a predator, seek out food or protect their young. Other species can pursue goals that, while not necessarily as complex as ours, are goals nonetheless. As persons we have the ability to be conscious of our goals, to articulate them and subject them to scrutiny. To ascribe goals to animals is not to

the majority (65.8%) of abortions were performed by ≤8 weeks' gestation, and nearly all (91.4%) were performed by ≤13 weeks' gestation. Few abortions (7.2%) were performed between 14–20 weeks' gestation or at ≥21 weeks' gestation (1.3%)" (Pazol et al. 2015). According to the American Congress of Obstetricians and Gynecologists, the second trimester ends at 27 weeks and six days (ACOG 2015). Thus on a sentience-based view, no more than 1.3% of fetuses aborted in the USA in 2012 would have had moral standing, depending on exactly which week of pregnancy the abortion in question occurred. Even if third-trimester fetuses have standing, this does not entail that aborting them is necessarily wrong. There could be cases in which the mother had such a strong interest in terminating her pregnancy, such as saving her own life, that it outweighed the interests of the sentient fetus.

6 My rationale takes the form of a principle rather than an appeal to a particular intuition or intuitions that are said to directly justify the DDE. Reviews of empirical research raise questions as to how widely such intuitions are shared. See May (2014: 5–7) and Feltz and May (2017).

ascribe to them any such ability to self-consciously apprehend or evaluate their purposes. It is to recognize their ability to form and act on purposes of their own, however limited those purposes may be compared to ours.

In Chapter 1 I characterized sentient beings as those that are conscious and have the capacity to experience pleasure and pain. I am unaware of any non-human species that matches this description but lacks the ability to form purposes. Mammals and other vertebrates have long been known to adjust their behavior to avoid pain and seek pleasure. This is only to be expected from an evolutionary point of view. It is not clear what evolutionary function would be served by the ability to experience pain without any corresponding ability to take steps to avoid it. The ability to form purposes however is conceptually distinct from sentience. It is possible to imagine a creature that is sentient but does not have the ability to form purposes of any kind. (My imaginary example of a "happy oyster" in Chapter 2 might be thought to fall into this category.) Such a creature could be harmed by being made to feel pain, but it could not be harmed in the manner with which the Doctrine of Double Effect is concerned. My version of the DDE will not apply to our dealings with such creatures, should any exist.

But while it is conceptually possible for an entity to be sentient and not be able to form purposes, in nature the two often go together. As much is suggested by the Cambridge Declaration on Consciousness, which summarizes contemporary science's understanding of animal consciousness:

> The absence of a neocortex does not appear to preclude an organism from experiencing affective states. Convergent evidence indicates that non-human animals have the neuroanatomical, neurochemical, and neurophysiological substrates of conscious states along with the capacity to exhibit intentional behaviors. Consequently, the weight of evidence indicates that humans are not unique in possessing the neurological substrates that generate consciousness. Non-human animals, including all mammals and birds, and many other creatures, including octopuses, also possess these neurological substrates. (Low et al. 2012)

The Declaration attributes to non-human animals "experienced feeling states, including those internal states that are rewarding and punishing" (Low et al. 2012). Although the Declaration does not refer to pleasure or pain, I take it for granted that they are among the rewarding and punishing mental states found in the animal kingdom. In attributing intentional behaviors to mammals, birds and other creatures, the Declaration recognizes such animals as also possessing purposes.

The Declaration has been criticized for stating the obvious (Bekoff 2012). Certainly we did not need to wait until 2012 to reasonably attribute consciousness to mammals and birds. What is less obvious, however, is just how widely sentience and the capacity to form purposes are distributed in the animal kingdom.

Consider the case of fish, which the Declaration does not mention. Recent research supports the view that they too are sentient. As one overview of fish cognition studies puts it, "gone (or at least obsolete) is the image of fish as drudging and dim-witted pea brains, driven largely by 'instinct,' with what little behavioral flexibility they possess being severely hampered by an infamous 'three-second memory'" (Laland et al. 2003: 199). Fish have nociceptors that transmit pain signals to the brain in a manner similar to that found in other vertebrates; the behavior of fish is thus negatively affected by painful stimuli (Sneddon 2003). The researchers who documented these findings subsequently injected the lips of rainbow trout with chemical stimuli (vinegar, bee venom) to which fish responded with increased breathing and reduced appetite, reactions similar to those of other vertebrates in pain. Additional research has revealed that administering morphine to trout that had been injected with noxious chemicals significantly diminished the fishes' pain-related reactions (Sneddon 2003; Sneddon and Leach 2016). In addition to pain, Chandroo et al. (2004) argue that fear and stress are evident as well. Such findings have been subsequently confirmed in follow-up studies and meta-analyses (Chandroo et al. 2004; Sneddon 2007; Braithwaite and Boulcott 2008; Nordgreen et al. 2009; Brown et al. 2011).

Findings of this kind have coincided with new discoveries regarding fish learning. A summary of 500 peer-reviewed articles on learning in fish for example found that they exhibit cultural traditions; co-operate to catch prey and examine predators; track the relationships of other fish; use tools; build nests; and possess social intelligence that allows them to pursue "Machiavellian strategies of manipulation, punishment and reconciliation" (Laland et al. 2003: 199). Such findings suggest that it is reasonable to attribute not only sentience but the capacity to form purposes to all vertebrate species.

Current research suggests that at least one non-vertebrate species should be similarly classified. The Cambridge Declaration attributes consciousness to octopuses. This is consistent with contemporary research that characterizes cephalopods has having evolved complex cognition in a separate evolutionary pathway from mammals (Godfrey-Smith 2013, 2016). They additionally possess a number of behaviors and faculties suggesting that they are conscious. They depend on learning visual clues to forage in their environments (Mather 2008), show evidence of play behavior (Kuba et al. 2006), possess the ability to recognize individual humans (Anderson et al. 2010), and the ability to discriminate between different escape routes from a maze, a form of complex learning (Hvorecny et al. 2007). When octopuses injure their arms they guard and favor the injured area for some time afterwards, a behavior normally thought to indicate pain (Alupay et al. 2014). The European Union includes octopuses and other cephalopods in its directive protecting animals used for scientific purposes, "as there is scientific evidence of their ability to experience pain, suffering, distress and lasting harm" (European Union 2010: 34). I include octopus but parsimoniously exclude, for the present at least, other cephalopods – such as cuttlefish, which Hvorecny et al. also show can learn conditionally – because the evidence to date is more convincing for octopuses.

Why should we grant any moral weight to animal purposes? To answer this question it is helpful to ask why we care about the purposes of human persons. In judging our interactions with other persons, we grant special weight to whether those interactions are consensual. There is a special horror associated with non-consensual sexual activity, for example. In ascribing moral importance to consent, it seems that we thereby ascribe moral importance to purposes. For what separates a consensual sexual encounter from a non-consensual one is whether or not someone purposely engaged in it. The capacity to form purposes seems closely related to our capacity for agency, something that we normally deem morally significant. Acting on our purposes is how we interact with the world, and whether we are able to carry out our purposes can have a significant impact on our well-being. In short it seems difficult to grant moral weight to consent, agency, well-being and other familiar concepts without also granting weight to purposes.

If we grant moral weight to the purposes of persons it seems only consistent to grant weight to those of merely sentient beings. This is true even though animals are not capable of consent as we normally think of it, which involves self-consciously agreeing to something. But for this fact our relation with animals seems similar to our relation to a fellow person who can give or withhold consent to our interactions. In both cases the entity with whom we interact is one for whom things can go better or worse; the entity is one that can engage in intentional behavior, acting on goals that significantly impact its well-being. Animals possess their own kind of agency, less complex than our own but a form of agency nonetheless, one that sees them want and do things. If thwarting the purposes of human beings seems at odds with respecting their agency, something similar seems true in the case of animals as well. In neither case do we say that every purpose commands moral respect. If a gunman or a bear threatened me, it would be permissible to resist. But all else equal, it seems we exhibit appropriate moral respect toward persons and merely sentient beings when we grant weight to their capacity for form and act upon purposes of their own.

The Animal-Friendly DDE

There is more than one way we might justify a version of the DDE that takes seriously the purposes of animals. Human persons can pursue purposes at the expense of animals' purposes, as when a hunter shoots and kills a deer when the animal is foraging for food. One possible ground to offer for a presumption against harmful direct agency directed at animals would thus be a prohibition on using them for human purposes. Let us call this the prohibition principle.

The Prohibition Principle: The presumption against harmful direct agency derives from the fact that such agency infringes an independent right not to be made to serve others' purposes.

This prohibition is too sweeping. My purposes can include petting a cat. This is the case whether or not the cat enjoys being petted. But insofar as the cat does enjoy being petted, my doing so also advances a purpose of the animal's. Given the possibility of acting in a manner that advances both a human and an animal purpose, there seems nothing necessarily wrong in the mere fact of using an animal for purposes of one's own.

This raises the possibility of a different rationale for the presumption against direct harmful agency. On this rationale, direct harmful agency toward animals is wrong when it occurs at the expense of the animal's own purposes. Let us call this the No-Expense Principle.

The No-Expense Principle: The presumption against harmful direct agency derives from the fact that such agency infringes an independent right not to be made to serve others' purposes at the expense of one's own purposes.

This, again, is too broad. There are many ways in which we interfere with an animal's ability to pursue its purposes. Suppose I am taking my dog for a walk in the park. He may see another dog that he wants to attack, but I prevent him from doing so, as my goals for the afternoon do not include witnessing a bloody dogfight. Alternatively I may be riding a horse that prefers to turn left but which I direct to turn right. There are countless ways in which human beings direct the behavior of animals. Of course many such ways are harmful, as when we direct them into slaughterhouses. The wrong in such instances, however, are captured by the concept of harm. The DDE by contrast, in the animal case as much as the human, is meant to include reference to a different moral wrong, one that involves a certain type of usage. Insofar as it is possible to use animals in gentle ways, as when we simply nudge them in one direction, a blanket prohibition on usage as such seems unwarranted.

There is however an additional possible rationale for an animal-friendly DDE. This rationale applies when we use an animal in such a way that we advance our purpose at the expense of the animal's very capacity to form purposes of its own. Let us call this the Capacity Principle.

The Capacity Principle: The presumption against harmful direct agency derives from the fact that such agency can infringe an independent right not to be made to serve others' purposes at the expense of the capacity to form purposes of one's own.

Suppose we reconceive trolley problems so that the individuals at risk of being killed are all human beings with a particular mental handicap. In the revised version of the trolley scenarios, all the potential human victims are merely sentient rather than persons. Now the choice is between killing one such human being to save five or letting five of them die. Would we give different answers to these trolley problems? Would we say that where we would not push the big man off the bridge in Wrestler, we would kill his merely sentient counterpart? I cannot see why we would change our answers. Everything is the same, but for the fact that the victims are merely sentient rather than persons. Surely that fact, by itself, should not change the morality of our response. To say this, however, is to say that the DDE should have a range of applications beyond persons to sentient beings. And if species is morally irrelevant, as I argued in Chapter 1, those sentient beings should include animals.

The version of the DDE based on the Capacity Principle can explain why this is the case. The prohibition on treatment the revised version articulates is rooted in respect for the capacity to form purposes of one's own. It does not seem arbitrary to grant moral weight to this capacity. There is something it is like to be an animal, and for all but the simplest forms of animal life, part of what it is like is the ability to do things and want things. A respect for the ability to form purposes is a respect for a fundamental aspect of vertebrate and octopus consciousness. It is the respect of not treating a sentient being "as a mere thing," and recognizing them as something more than "material to be strategically shaped or framed by [one's] agency" (Nelkin and Rickless 2014: 154; Quinn 1989: 348).

This rational for the DDE does not prohibit using animals. Insofar as usage of an animal can be gentle or harmless, the present principle does not counsel against it. It rather issues a prohibition against lethal forms of usage, those that destroy a sentient being's ability to form its own purposes by killing it. It is this type of usage that the revised DDE categorizes as a second distinct wrong, beyond that of harm. When someone harms sentient beings in this particular way, he "takes up a distinctive attitude toward his victims. He must treat them as if they were then and there for his purposes" (Quinn 1989: 348). The animal-friendly DDE is similar to the DDE-R in identifying this attitude as morally wrong.[7] It is different insofar as it is concerned with instances in which this attitude leads to the death of the animal. The person-directed California version applies to direct harmful agency as such.

We might wonder about this difference. After all, there would seem to be a large range of harms that are the result of direct agency that stop short of death. I might for example cut off an animal's limb in a way that does not kill it. At the lower end would be temporarily causing an animal pain or distress to serve some end of my own. The animal-friendly DDE as I have outlined it has no bearing on these cases. It confines itself strictly to lethal instances of direct agency. Hence the question: why restrict the domain of the animal principle's application in this way? Why not appeal to a more general principle that would apply to non-lethal forms of usage? Such a principle might take the following form:

> The Weighted No-Expense Principle: The presumption against harmful direct agency derives from the fact that such agency infringes an independent right not to be made to serve others' purposes at the expense of one's own purposes, when one has a highly ranked interest in pursuing one's own purpose.

Call this the Weighted principle for short. It avoids the shortcomings of the first two principles canvassed. It is rooted in respect for an animal's most fundamental purposes. As such, it will have a wider application than the Capacity principle. In this and other ways it is a highly attractive

[7] My view is patient-centered rather than agent centered. For the distinction see Nelkin and Rickless (2014).

principle. So why not ground a sentience-sensitive DDE in it rather than the capacity principle?

The simple answer is that my argument is deliberately conservative. Making the case for the Weighted principle would require a detailed outline of what makes one interest more important than another. This raises the possibility of critical responses that offer their own alternative criterion of what makes one interest more important than another. For the purposes of critically analyzing the new omnivorism, this debate seems unnecessary and worth avoiding. If anything counts as a fundamental interest it is the interest in staying alive, which is a perquisite of advancing any interest at all. Additionally, simply getting clear on the DDE as it applies to persons is already difficult enough, even with decades of philosophical debate over ever-more-refined principles. Non-anthropocentric forms of the principle are only now beginning to receive serious consideration. It will therefore be progress enough to demonstrate that *lethal* direct agency is especially morally wrong, whatever else proves to be the case. I therefore rest my case with the Capacity principle and do not seek to make the case here for the Weighted principle.

Broadening The Kantian Rationale

A new version of the DDE requires a new rationale. I believe such a justification is already pregnant within the Kantian rationale invoked by California Deontologists. Quinn notes that if harmful direct agency is worse than its indirect counterpart, it is because it involves a wrong over and above simply harming or not preventing harm. As he writes, "harmful direct agency must have two things against it, while equally harmful indirect agency need have only one" (1989: 348). Two strands run through Quinn's analysis of the special wrong of direct harmful agency. One of these strands focuses on considerations that would extend the DDE to sentient animals while the other highlights factors that would make the DDE irrelevant to our dealings with sentient animals. The fact that the inclusive considerations are more compelling provides the first of several reasons to apply some version of the DDE to sentient animals.

Quinn suggests that the additional wrong present in harmful direct agency concerns the particular way in which victims feature in an agent's thinking. In cases of indirect harmful agency, we can be sure that our actions will have victims. But in indirect cases their involvement in our action is not useful to our end. An agent of direct harm, however, "has something in mind for his victims – he proposes to involve them in some circumstances that will be useful to him precisely because it involves them. He sees them as material to be strategically shaped or framed by his agency." (348) An agent of direct harm thus exhibits a particular attitude toward his victims. The agent treats them as if they existed for his purposes. This, however, violates an important moral presumption: "Those who can be usefully involved in the promotion of a goal only at the cost of something protected by their independent moral rights (such as their life, their bodily integrity, or their freedom) ought, prima facie, to serve the goal only voluntarily" (349).

Quinn notes that there can be exceptions to this rule: cases in which someone is obliged to give herself over to the pursuit of a goal even at great personal cost. Cases of this kind might include conscription or paying taxes. But in the trolley and other cases I have described, the victim is made to play a role that is not required of them. It is this element that generates the second wrong present in cases of direct harmful agency, above and beyond the harm. This second element is clearest in instances of what Quinn terms opportunistic agency, of which both Strategic Bomber and Wrestler are examples. In such cases the victim's presence is an opportunity to be exploited by the agent along the way to achieving his goal. This is different from what Quinn calls eliminative agency, in which the victim's presence is not an opportunity but an obstacle. When Craniotomy is approached in a Catholic framework, for example, the presence of the fetus is a problem that is overcome by killing it, which is an act of eliminative agency. Quinn's point is that the DDE more strongly condemns direct harmful *opportunistic* agency than direct harmful eliminative agency. While both are worse than indirect agency, opportunistic agency involves the ongoing use of an unwilling victim. Quinn suggests that this special heightened censure of opportunistic harm explains why especially severe forms of opportunistic agency, such as torture, are impossible to justify.

Some aspects of Quinn's Kantian rationale cannot be easily extended to apply to animals. Insofar as animals are incapable of consenting to anything, a rationale based on consent is inapplicable to them. Some forms of harm to animals also cut across Quinn's distinction between opportunistic and eliminative agency. Often it is an animal's material body, its flesh or its skin, that is viewed as a resource. Insofar as the animal is alive it is an obstacle, but once it is dead it becomes a thing of value. The animal's mind is thus eliminated in order that its body can be used opportunistically. This suggests a third category between opportunistic and eliminative harm that combines aspects of both.

Many other aspects of Quinn's Kantian rationale however do seem to extend to sentient animals. It is certainly possible for example for an agent to involve them in some scheme that will be useful to him precisely because it involves them. Similarly, it is possible to view animals as mere material to be shaped or framed by one's agency. In his discussion of independent moral rights, Quinn mentions as candidates for such rights life, bodily integrity and freedom, all attributes that animals naturally possess. Finally, if we think of the entire range of harms it is possible to inflict on animals, torture would seem especially difficult to justify in their case also.

Despite the many animal-inclusive criteria in Quinn's rationale, it is presented in specifically human terms:

> The doctrine reflects a Kantian ideal of human community and interaction. Each person is to be treated, so far as possible, as existing only for purposes that he can share. This ideal is given one natural expression in the language of rights. People have a strong prima facie right not to be sacrificed in strategic roles over which they have no say. They have a right not to be pressed, in apparent violation of their prior rights, into the service of other people's purposes. (350–1)

A Kantian ideal of human community is a community of human persons. As persons we are rights-holders, and in such capacity we have the dependent right not to be presses into serving the purposes of others. We have a "veto power" over attempts to use us in this way, a power that reflects our status as "free and equal agents" (351). Depending on how personhood is defined there may be some animals that are persons. But traditional definitions of personhood normally exclude beings that are merely sentient rather than self-conscious. Quinn's reference to "free and equal" agents suggests such a view of personhood. The equality that deontology emphasizes is an equality of beings able to employ reason and moral deliberation, among other attributes confined to persons. A traditional view of personhood is further suggested by Quinn's discussion of whether or not agents of direct harm "show a shocking failure of respect for the persons who are harmed: they treat their victims as they would treat laboratory animals" (348). This passage occurs in a discussion of a rational for the DDE that Quinn ultimately finds inadequate. Nevertheless, the use of laboratory animals as a contrast class clearly excluded them from the category of persons. They and other animals fall outside the community of free and equal agents, and so are not covered by Quinn's version of the DDE.

Quinn's post-Kantian understanding of the moral community and a community of human persons faces a familiar problem. It would not seem to have the resources to provide an appropriate place for human beings who are not persons, such as those born with certain severe mental disabilities. We could for example create such human beings and use them the same way we do laboratory animals. Why do this in the case of rats but not severely mentally handicapped human orphans? Quinn's account, like so many others, does not seem able to provide a cogent rationale.

Our concern here however is not the argument from species overlap but the grounds on which to exclude merely sentient beings from the protection of the DDE.[8] To that end I believe the key consideration on Quinn's account concerns the applicability of rights. In emphasizing the status of persons as rights-holders, Quinn draws a distinction with the notion of interests. However, when we bring critical scrutiny to bear on Quinn's distinction between rights and interests, it becomes clear that he overstates their conceptual separateness. Indeed, Quinn's own account presupposes an appeal to the significance of interests. As such, it becomes clear that animals, as entities with morally relevant interests, warrant the protection of the DDE.

The reason that Quinn ultimately rejects the idea that the DDE is based on a failure to exhibit respect to one's victims is that such disrespect can also be found in cases of indirect harmful agency. A strategic bomber, for example, could kill a group of civilians as part of an operation that destroys an unimportant target. Although the bomber does not run afoul of the DDE, he has nevertheless failed to "treat his victims with the consideration that they and their interests deserve" (348). On Quinn's account, the particular form of moral disrespect the DDE rules out is a failure to respect rights rather than interests. If rights were exclusive to persons, this would naturally explain why animals are excluded from the DDE.

This grounds for denying the protection of the DDE to merely sentient beings presupposes that there is a strong difference between interests and rights. Whether this is the case, however, will depend on how we theorize rights. On one leading theory of rights, for example, that of Joseph Raz, rights are ultimately grounded in interests. As Raz famously defines rights, "'X has

8 The argument from species overlap is a better term for the so-called argument from marginal cases. Among other advantages, it does not label some human beings as marginal. For discussion, see Horta (2014).

a right' if and only if X can have rights, and, other things being equal, an aspect of X's well-being (his interest) is a sufficient reason for holding some other person(s) to be under a duty" (Raz 1986: 166). On this account rights are not a bedrock moral concept, but are rather grounded in interests. Sentient beings may not have all of the rights we have, but it is not clear why they do not have some interests that ground at least some rights (Cochrane 2013: 660). If so, the reason for excluding merely sentient beings from the category of entities entitled to the protection of the DDE evaporates. Insofar as sentient beings can have interest, such as an interest in avoiding suffering, they meet the condition necessary to possess rights on the interest-based theory of rights. As such, there will be no grounds on which to deny them the protection of the DDE.

Of course it is possible to deny that rights are in fact grounded in interests. An alternative approach is to ground them in personhood. This approach however faces problems both familiar and debilitating. For one, it is not able to extend rights to all human beings. Those who are not or will never become persons are denied rights, an outcome that is at odds with how human rights are normally conceived. Alternatively, proponents of personhood-based theories of rights have sometimes appealed to *potential* personhood as grounds to extend rights beyond existing persons. But this still fail to extend rights to all human beings as it is simply not the case that all members of our species will become persons. In addition, as Alasdair Cochrane notes, potentiality arguments are "notoriously problematic ... it would be absurd to grant me the rights of a supreme court judge on the grounds that I have the potential to become one at some point in my future" (2013: 662).

A separate strategy for extending a person-based account of rights to all humans has been to argue that human non-persons warrant rights because they belong to a group of whom it is typical to have rights. Often, that group is the species *Homo sapiens*. But this approach faces the same problem as the previous strategy. It would be absurd for me to argue that although I did not hand in my homework I nevertheless deserve an A on the grounds that I belong to a group of students of whom it is typical that they received As. The strategy also faces the separate problem of justifying species membership as a morally relevant group. What makes this category more salient, for example, than that of sentient beings? (Nobis 2004). As we saw in Chapter 1, Arguments for this view fail to justify species as a morally relevant group.

Agricultural Ethics and the DDE

We are now in a position to see the relevance of the animal-friendly DDE. If sound, it entails that in comparing the harm different agricultural regimes do to animals, there is a morally relevant consideration beyond the sheer number of animals killed. There is also the question of whether or not their deaths were the result of direct or indirect harmful agency.

Recall that according to Steven Davis, his preferred agricultural model was estimated to kill 1.35 billion animals per year, in comparison to the 1.8 billion he suggested would be killed by a vegan model. Let us suppose for the sake of argument that these numbers are correct. Even so, there is the separate consideration that animals in the Davis model will be the victims of direct harmful agency while those in the vegan model will be victims of indirect harmful agency. As such, there is a separate wrong present in Davis's preferred model. In Wrestler, the death of the eponymous athlete is considered unjustified even though killing him would save five lives. This raises the possibility that the vegan model would still be justified even if it killed *five times* the animals as did Davis's model. If a multiplier of five does not suffice to justify harms in the human case, it is not clear why less than five should suffice to justify direct harm in the sentient animal case. At a minimum the burden of proof seems to reside with someone who would argue that a multiplier of five suffices to justify harmful direct agency directed at merely sentient beings but not persons. Until that burden is met, even granting Davis's numbers, the animal-friendly DDE provides an avenue of escape from his conclusion.

The animal-friendly DDE has bearing on other argument for burger veganism. Consider in this regard a powerful case for hunting made by Lawrence Cahoone. He cites the fact that plant-based agriculture harms animals in at least five ways: by clearing their habitat; through the use of pesticides and fertilizers that pollute their water; by employing machinery that runs them over; by killing wildlife that might feed on crops; and through indirect harms caused by agricultural machinery and methods that are "processed and refrigerated with power from river-altering, coal-burning or nuclear-waste-producing plants" (Cahoone 2009: 79). When we weigh up these harms against those done to animals by hunters, he argues, veganism entails some form of hunting:

> hunting must be judged not only relative to the animals hunting kills, but relative to the *animal cost of the agrarian activity that would replace the meat from the animals it kills*. Where a type of hunting has a lower death/nutrition ratio than a type of farming, and where the pain of death by hunter is

> arguably no greater than death by farmer, the meliorist version of the animal welfare/animal rights argument must morally prefer hunting to farming. (Cahoon 2009: 80, emphasis in original)

There is truth in what Cahoon says, but it is overstated. With the exception of killing wildlife that might feed on crops, all of the harms Cahoone associates with plant-based agriculture are indirect. Thus even if hunting kills fewer animals than those that are indirectly killed by farming, bracketing for the moment direct deaths due to farming, this will not suffice to show that it is morally superior. Hunting is clearly a form of direct harmful agency. As such it must meet a higher standard of justification than the indirect harms caused by crop cultivation.

Cahoone anticipates this objection. He does so in a passage that uses a variation of strategic bombing as an example. It involves bombing a factory even though it kills children at a school next door:

> in just war theory that doctrine [of double effect] was supplemented by others, notably proportionality, which rules out disproportionate harm whether or not it is purposively sought. So the anticipated deaths of the children, *not* immoral under double-effect, could still make the bombing immoral if the total harm caused is disproportionately high relative to the good it achieves (e.g. hastening an end to the war). Double-effect by itself may not be salutary in animal-treatment or environmental ethics. (Cahoone 2009: 86)

What Cahoone says here is true. The distinction between harmful direct and indirect agency is not the only consideration we need to bring to bear on strategic bombing and other cases. We also need to take into account proportionality. But Cahoone fails to adequately note the ramification of double effect "by itself." It is enough to show the inadequacy of his appeal to hunting's "lower death/nutrition ratio." Insofar as hunting violates the animal-friendly DDE, it will be worse than any farming practices that, all else equal, kills an equivalent number of animals without violating the DDE, and worse than any farming practices that, all else equal, kill even more animals without violating the DDE.

This still leaves the deaths of wildlife caused to protect crops. The animal-friendly DDE as I have outlined it clearly applies to these animals. For this reason Cahoone is right to note that there could be cases in which a diet containing meat from hunted animals is to be preferred on protectionist grounds to a diet containing plants that were cultivated in a way that required intentionally killing wildlife. Cahoone however is wrong to conclude that, "*it is very likely that agricultural production kills more animals than deer hunting per unit of nutrition*" (81, emphasis in original). Cahoone bases this conclusion on Davis's calculations regarding field animal deaths, which as we saw are mistaken.

Is there a protectionist case for hunting that does not rely on Davis's numbers? On a protectionist account, a superior alternative to vegan hunting would be eating plants cultivate in a manner that required killing no wildlife. As we saw in Chapter 3, this already occurs when fruits and vegetables are raised in greenhouses and indoor farms. Vegan hunting would additionally be second best to eating crops that were protected with non-lethal forms of wildlife control. Protectionism will encourage both of these alternatives before hunting. A protectionist rationale for hunting would also appear to face one of the same problems as venison veganism, namely, that such a diet may be more difficult to follow than one that abides by a strict no-meat rule. Cahoone's argument is meant to justify eating wild but not domesticated animals. Practitioners of such a diet would need to avoid the risk of self-deception and weakness of will posed by behavioral rules that require interpretation. Consideration raised by Cahoone therefore do not suggest that protectionists should condone hunting, as opposed to viewing it as better than agricultural practices that, all else being equal, involve greater direct harm to animals.

In order for protectionism to endorse hunting we would need to know its full costs, as they involve both direct and indirect harm. Cahoone, for example, mentions the deaths of animals due to the power required for refrigeration or being run over. By that standard, harm to animals that hunters caused by refrigerating their carcasses or accidentally running them over on hunting trips would also have to be counted. And if the collateral deaths of field animals are to be included in our moral calculus, then so should the collateral deaths of human beings.

Under a protectionist rationale, vegan hunting would be the only ethically acceptable way for people to eat meat. It is natural to think this would cause more people to take up hunting. As it is currently practiced hunting is a safe pastime. In 2012 for example, 1,227,025 hunting licenses were sold in Texas. During the same year only three people died in hunting related accidents, a rate of 0.2 fatalities per 100,000 licenses (Texas Parks and Wildlife 2013). But if hunting grounds contained a greater density of hunters, as a Cahoone-style scheme would encourage, the fatality rate might increase. Widespread vegan hunting could also see a greater number of firearms in people's homes, which have been associated with a higher risk of firearm suicide and firearm homicide (Dahlberg et al. 2004).

The more widely vegan hunting were practiced, in short, the more it is likely to have negative outcomes that will call its advisability into question.

Given these considerations, while it is possible to imagine hypothetical circumstances in which hunting was permissible or even obligatory on a protectionist account, such circumstances are unlikely to arise in practice. Even in a world in which hunting did result in less direct harm to animals, we would still be wronging them in a manner the animal-friendly DDE singles out for condemnation. That is, even if hunting were the only way to avoid directly wronging some larger number of animals, composed of those killed to protect crops, vegan hunting would not be morally good. It would rather be morally tragic. The case for hunting would be that it was the lesser evil. Rather than accept this as the best we can do, the animal DDE and protection theory more broadly compel us to aim for an outcome that avoids evil altogether.[9]

The title of Cahoon's paper is "Hunting as a Moral Good." From the point of view of a version of protectionism that employs the animal-friendly DDE, this conclusion is open to doubt. Cahoon presents no evidence to suggest hunting reduces the amount of direct harm to animals. Given that hunting clearly does involve directly harming animals, it is reasonable to withhold endorsing vegan hunting until evidence for its harm-reducing status is presented.

Possible Objections

This ends my outline and application of an animal-friendly DDE. In endorsing an animal-friendly version of the DDE, I face a possible objection that can be pressed against any version of the DDE. That objection concerns a variant of the trolley problem known as the Loop Case. This case is identical to Lever in every way but one. In Loop, the side track down which one can divert the trolley loops back to the main track. As a result, if the one person were not on the side track, diverting the trolley would not in fact save the five, because the trolley will simply loop back to the main track and run them over. This version of the problem thus

[9] In a similar way, we should not be satisfied with defenses of meat eating that stress how painful it is for animals to be run over by farm machinery, in comparison to the relatively painless way in which they are killed in humane slaughter systems. Even granting this empirical claim, it still leaves unaddressed the special wrong of deliberately killing them for food. In addition, there is the fact that domesticated animals have their natural functions frustrated by confinement while field animals do not. It is therefore not clear that when the animals' lives are judged in their entirety, the fate of the domestic animal is in fact superior to that of its free-living counterpart.

involves using the one person: it is by fatally diverting the trolley into him or her that the five are saved. In this way, it is identical to the Wrestler case. But a common reaction to the Loop case is to say that it is permissible to send the trolley down the looping sidetrack. This is a problem for any version of the DDE, as a consistent application of the DDE would seem to prohibit diverting the trolley down the sidetrack. But the notion that it could be wrong to divert in Loop but not in Lever has struck some philosophers as incredible. As Judith Jarvis Thomson, the inventor of the Loop case has put it, "we cannot really suppose that the presence or absence of that extra bit of track [that loops back] makes a major moral difference as to what the agent may do in these cases (Thomson 1985: 1403).

Any account that takes the DDE seriously must disagree with Thomson. The presence of the looping track transforms the nature of the death of the one person. As such, it becomes impermissible to use them as a means of stopping the trolley. Such a view will strike some as counterintuitive. Nevertheless, we should not accept this as grounds to abandon the DDE. One reason is because any theoretical account of our moral obligations is likely to have some counterintuitive outcomes. It is not clear that a philosophical view with no such implications is possible. In addition, there is a cost to affirming that it is permissible to divert the trolley in Loop. It entails the permissibility of not only killing the wrestler, but diverting the Trolley in a wide range of similar cases. Consider in this regard the Loop Bridge Case, which resembles Wrestler in that there is again a bridge over the track on which a person is standing.

The Loop Bridge Case: Your pushing a lever causes the trolley to be diverted onto a perfectly circular track that loops back around to the same point where the trolley was on the main track before it was diverted, whereupon it proceeds in the direction of the five after this brief circular detour. Let us suppose, further, that the trolley's travelling in this circle remotely triggers [a] pole on the bridge over the main track to move . . . thereby causing the person on the bridge to fall onto the main track and into the path of the trolley after it has circled around the side track and resumed its travel down the main track. His being hit will stop the trolley and kill him. (Otsuka 2008: 107)

In the Loop Bridge case, the death of the person on the bridge is the means by which the lives of the five are saved. In this way, it resembles both the original Wrestler case and the Loop case. Now consider the Ramp Case:

The Ramp Case: You can stop the trolley from hitting the five by opening a drawbridge. So doing will send the trolley up the ramp of the open drawbridge and into the air, where it will hit one person standing on a pedestrian bridge overhead. The trolley's hitting the one will stop its forward progression and

> cause it to fall into the water below the drawbridge. The trolley's hitting the one will also kill him. If, however, the one had not been overhead, then the trolley would have managed to fly over the gap, land on the downward sloping ramp, and hit and kill the five. (Otsuka 2008: 107–8)

In the Ramp case, the death of the person on the bridge is the means by which the lives of the five are saved. In this way it too resembles both the original Wrestler case and the Loop case.

Michael Otsuka has suggested that killing the one in Loop-Bridge and Ramp cases is counterintuitive. Anyone who shares this reaction (as I do) will have to let go of this intuition in accepting the permissibility of diverting the Trolley in Loop (Otsuka 2008: 110). But there is no morally relevant difference between Loop, Loop Bridge, Ramp and Wrestler. Here the presence of extra looping track, like that of the ramp, really is irrelevant. The popularity of the view that it is permissible to divert the trolley in Loop may therefore be an artefact of presenting Loop without thinking through the ramifications for a full range of cases. So far as the DDE is concerned however, it is impermissible to kill the one in Wrestler, Loop, Loop-bridge and Ramp alike.

A separate possible objection to my account is that the trolley scenarios are disanalogous to the agricultural scenario. In our capacity as food consumers, we direct resources to the farming industry. In this way our financial choices feature in the causal story of how animals die in agriculture, whether in slaughterhouses or the fields, regardless of which particular farmer our money goes to. For without our financial support, there would be no agricultural industry in the first place. The trolley scenarios are different, on this view, in that we did not put the tracks or bridge or wrestler there: the basic elements of the scenarios are assumed to have come into being independent of our actions. Given this difference, we can't derive moral lessons from trolley cases that apply to choices involving agricultural.[10]

We can see the problem with this objection by considering a revised version of Wrestler. Imagine that the bridge he is standing on is public property. It has been funded with tax revenue, and the person who must decide whether to push the wrestler to his death is a taxpayer. Do these new details have bearing on the person's decision? It is hard to see how. The agent is still faced with the same choice of letting five people die or killing one person. Nothing in the new backstory about how the bridge

[10] I am grateful to Valéry Giroux for this objection.

came to be changes the central fact that the person must decide whether five people or one will die. Similarly, even if we are complicit in the existence of modern agriculture, this does not change the fact that we are faced with the decision whether or not to buy meat or plant food. It does not seem right to hold that, given our previous patronage of the industry, no special wrong occurs when we opt for animals to be deliberately killed for food.

It is also unclear how much choice we have when it comes to patronizing the agricultural industry. While in principle it might be possible to grow all our own food, buying food is something we are socialized into from a young age, and the modern economy is defined by a division of labor that makes it impractical to devote many hours to harvesting our own food.

The DDE and Infanticide

The version of the DDE defended here has bearing on the issue of life-saving infanticide. In Chapter 2 we noted the case of an infant with no family of any kind who could potentially be killed to provide organs that would save the lives of three toddlers. In outlining such a scenario I offered why such a practice should be illegal even if it is moral. But is it moral? If the protection provided by the DDE extends to merely sentient beings, then this calls into question the morality of deliberately killing an infant in order to save three children.

The expanded DDE says that sacrificing the life of one merely sentient being requires greater justification than saving the lives of five merely sentient beings. The life-saving infanticide case is more complex insofar as the lives to be saved are those of three persons. Are three persons worth the life of five merely sentient beings in this context? I am not sure how to answer this question. The expanded DDE, however, should cause us to realize that even if the lives of three persons ultimately matter more than the life of one merely sentient being, whether an animal or an infant, this fact by itself does not automatically justify sacrificing the merely sentient being to save the others. The case for life-saving infanticide however took as a key part of its rationale the fact that the life of one infant is less significant than that of three children. If the expanded DDE is sound this comparative weighting could be true yet still not enough to justify deliberately killing the infant. On such a view there could still be cases in which it was permissible to kill an orphaned infant to save the lives of persons, but the case will be clearest when some very large number of persons' lives are at stake, which one hopes will be a condition unmet outside of wartime and other extreme circumstances.

Insofar as there is any doubt that we should kill a sentient being, whether animal or human, it seems reasonable to want clear and unambiguous evidence that a high threshold of justification has been met. If it is doubtful that threshold is met in the life-saving infanticide and similar cases, this provides grounds to doubt that we should kill the infant, even if infants do not possess primitive self-consciousness (a possibility I explore in Chapter 6). These grounds for doubt afford a more purely moral reason to support the universal legal ban on life-saving infanticide discussed in Chapter 2.

Conclusion: The Ecumenical DDE

My argument for an animal-friendly DDE is in keeping with recent efforts on the part of deontologists to address their doctrine's traditional blind spot regarding animals. On a traditional Kantian view, animals are denied moral standing because they lack the ability to scrutinize the grounds of belief and action, and therefore the ability to recognize or formulate moral law. On Kant's view, it is only possible to have moral obligations to other rational beings that are "ends in themselves" in this sense. Christine Korsgaard is a prominent modern deontologist who argues that this is a mistake. While not being ends in themselves is a legitimate reason to hold that animals cannot have moral obligations *to us*, Korsgaard writes, it is not grounds to conclude that we have no obligations to them.

> *Among* the interests we protect through our own moral legislation are interests that spring from our animal nature, not from our rational nature. Our objection to pain and suffering is an obvious example . . . . So although being rational autonomous beings – moral lawmakers – is what enables us to *make* normative claims on ourselves and each other, the content of those clams cannot be given completely by respect for rational autonomy itself. (2005: 108)

California deontology is not synonymous with Kantian deontology. Nevertheless, its view of the DDE as only applicable to beings capable of consent inherits the exclusionary aspect of traditional deontology. Korsgaard's appeal to the moral relevance of interests rooted in our animal nature, which is a nature we share with other animals, is in keeping with the recognition of the moral interests of animals that justifies the animal-friendly DDE. Insofar as many other contemporary deontologists also recognize the moral significance of animal interests, it seems reasonable to view the animal-friendly DDE as broadly compatible with some form of deontology (Franklin 2004; Kriegel 2013).

Some proponents of burger veganism employ frameworks incompatible with endorsing the animal-friendly DDE. George Schedler, for example, makes such a case in a utilitarian framework, which has traditionally been seen as opposed to the DDE (Singer 1980; Schedler 2005; Singer 2011: 183). This might be thought to limit the relevance of the animal-friendly DDE in the debate over burger veganism. Insofar as Schedler and other critics mount arguments incompatible with any form of deontology, can they dismiss arguments that appeal to the DDE? When we recall the wide variety of moral theories invoked in the debate, the value of a specifically deontological device might seem limited indeed.

Recently however there has been a move on the part of consequentialists to "consequentialize" seemingly opposed moral theories. According to Douglas Portmore, this approach involves taking

> whatever considerations that the non-consequentialist theory holds to be relevant to determining the deontic statuses of actions and insist that those considerations are relevant to determining the proper ranking of outcomes. In this way, the consequentialist can produce an ordering of outcomes that when combined with her criterion of rightness yields the same set of deontic verdicts that the non-consequentialist theory yields. (2007: 39)

A consequentialized version of the doctrine of double effect would seem more easily assimilated into rule-consequentialism than act consequentialism. And such a principle would be importantly different from a deontological version of DDE. The point however is that the DDE is in principle compatible with more than one ethical framework. This has long been evident of traditional conceptions of the DDE, which have found a home in both deontology and Catholic natural law philosophy. It seems at least an open possibility that the animal-friendly DDE might find a home in both these frameworks as well as within some form of rule-consequentialism, virtue ethics and common-sense morality. If so it would make the animal-friendly DDE relevant not only to one particular approach to ethics, but a wide variety of influential frameworks. This would mean that a proponent of burger veganism would not be able to refute it simply by attacking one particular background theory.

Of course, critic such as Schedler can still insist on mounting their case from within a particular framework that is incompatible with the animal-friendly DDE. Act-consequentialism would seem the most prominent example. But a critic who adopts this strategy must take on the thankless task of refuting the many well-known criticisms of act-consequentialism. If it turns out to be the case that the animal-friendly DDE is not tethered to one particular moral framework, but is instead compatible with a plurality of them, this will prove an especially attractive feature.

CHAPTER 5

Killing Them Softly

Introduction

Animal protection theory and animal agriculture have traditionally been seen as representing opposing values. The reasons for this are not far to seek. When Peter Singer published *Animal Liberation*, for example, it contained a graphic chapter devoted to exposing the overlooked suffering of food animals, a consideration that motived more than one reader to stop eating animals. Since the turn of the twenty-first century, however, the relationship between protection theory and animal agriculture has become more complicated. In 1997 the McDonald's corporation agreed to improve its animal welfare standards following a long campaign by animal activist Henry Spira (Singer 1998: 166–77). After the company's implementation proved to be slow and reluctant, the animal rights group People for the Ethical Treatment of Animals (PETA) launched an anti-McDonald's campaign. Eleven months later the company pledged to start auditing its meat suppliers and to only work with those that used Temple Grandin's humane slaughter systems. Following additional lobbying by PETA, Burger King and Wendy's introduced similar reforms, which have been characterized as having a ripple effect throughout the industry (Grandin 2001: 106). Singer and his work played a causal role in this outcome. Not only was Spira a former student of Singer's, the founder of PETA, Ingrid Newkirk, launched her group after reading Singer's book. As for Singer himself, he was involved in Spira's campaign and attended negotiations with McDonald's (Singer 1998: 172–6). The contemporary slaughter industry, therefore, rather than fundamentally opposing protectionist values, has been partly shaped by them. The industry today publicizes its commitment to animal welfare (McDonald's 2018). In this and other ways it wishes to be seen to take to animal suffering seriously, in a manner that was not true when Singer and others took up the moral status of animals in the 1970s.

The rise of humane slaughter and Singer's support for it have given rise to the view that Singer's philosophy of animal liberation, rather than requiring veganism, entails only eliminating factory farmed meat from one's diet. Proponents of this view cite passages in Singer's work that suggest killing animals can be permissible when it is done painlessly, among other conditions. This reading of Singer has been put forward during the same historical moment when humane slaughter has not only been adopted on a wide scale but has been widely characterized as making possible the ethical slaughter of food animals and so permitting "eating meat with an easy conscience," as one newspaper headline puts it (Irish Independent: 2009). As it is expressed in the popular press and elsewhere this view of humane slaughter is often phrased in speciesist terms. Certainly few of the companies that slaughter animals or consumers who buy meat from them are proponents of animal rights. What lends significance to the suggestion that Singer's work justifies humane slaughter is that if this reading is correct, it means there is an influential strand of protection theory that comes to the same accommodating conclusion as the increasingly popular speciesist view of humane slaughter. This reading of Singer, in short, raises the possibility that it is possible to endorse anti-speciesism and the equal consideration of interests and continue to eat other species. We can have our meat and eat it too.

What follows is an evaluation of the claim that humane slaughter is the arrangement most in keeping with Singer's anti-speciesist philosophy, a reading that has been advanced most prominently by Gary Francione. Grandin's system as it currently exists permits some animal suffering. I examine this real-world version of the system particularly as it applies to chicken, the species that is by far killed in the most numbers. I also examine a hypothetical ideal version that eliminates all suffering from the slaughter process. In doing so, I distinguish two versions of Singer's argument for animal liberation, one based on utilitarianism and the other on equal consideration. When it comes to Grandin's existing system, the fact that it still permits some suffering makes it morally wrong according to both of Singer's arguments. Both arguments would permit an idealized version of humane slaughter in which the system really was painless and all the animals were merely sentient. In the case of Singer's utilitarian argument, however, such justification comes at the price of rendering persons morally replaceable, among other unattractive implications. Although Singer's equal-consideration argument avoids these implications, it too ultimately fails to justify an idealized version of humane slaughter. Among other reasons, the equal consideration argument fails to recognize that death harms even merely sentient beings by frustrating their interest in continued existence.

When it comes to humane slaughter as currently practiced, Singer's arguments correctly understood both deem it a lesser evil to traditional slaughter, not a morally justified practice in its own right. Such a view is consistent with a wide range of animal protection theories, including those that, unlike Singer's, embrace the idea of animal rights. As such there is no special link between Singer's views and humane slaughter as currently practiced. The claim that his view exonerates humane slaughter systems such as Grandin's fails to succeed.

Grandin's Method of Slaughter

Temple Grandin is well known as a successful and prominent person with autism. Oliver Sacks' 1995 book, *An Anthropologist on Mars*, was titled after a phrase Grandin used to describe her life as a person with autism in a non-autistic world. Grandin has since gone on to become a bestselling author in her own right, and writings by and about her highlight her career as a designer of humane slaughterhouses. Systems designed by Grandin now handle half the cattle killed for food in Canada and the United States (Grandin 2001: 103). Facilities she designed are also located in Europe, Asia, Australia, and South America. Chances are good that any readers of this work who eat meat will have consumed beef or pork processed in a Grandin plant at least once.[1]

Grandin has written that much of her success in working with animals comes from the fact that "I see all kinds of connections between their behavior and certain autistic behaviors" (2006a: 172).[2] She gives the example of high-pitched noise. Just as someone whistling in the middle of the night will cause her heart to race more than it would that of a non-autistic person, animals are easily startled by noises such as a bell or the sudden hiss of an air brake (2006a: 169). Grandin's system therefore not only minimizes high-pitch sounds that animals can hear, but also eliminates many visual details that loom large from an animal's point of view. In the case of cows, for example, an entire herd can

[1] For a small sampling of Grandin's electronic media coverage, see Grandin 2017. Typing Grandin's name into the Factiva newspaper database calls up over 1,600 articles from across the English-speaking world.

[2] Karen Davis has challenged Grandin's claim that her system of slaughter is inspired by her autism. "Many of the problems Grandin presents herself as uniquely spotting in the slaughterhouse environment are the kinds of things that an intelligent non-autistic sees on entering an inbred culture" (Davis 2005: 1). Grandin's emphasis on a link between autism and animal behavior is noticeably more pronounced in her popular books than in her academic writings and may sometimes be slightly exaggerated. However, I am more inclined to accept it than Davis is. Among other reasons, there have been cases of other autistic people identifying strongly with animal behaviors (e.g., Prince-Hughes 2004).

stop if it comes across a swinging chain, which will cause the lead cow to move its head back and forth with its swing. Similarly, strong visual contrasts such as shadows, light reflecting in a puddle, or a drain running across the animals' path will cause balking. Even something as seemingly minor as a styrofoam cup on the ground or a piece of cloth flapping in the wind can cause a herd to freeze up (2006a: 167–8). Grandin's system meticulously eliminates all such distractions, which can cause the animals to stop moving.

Grandin has written that "I was always a cow and pig person, not a chicken person" (Grandin and Johnson 2010: 207). By this she means that she has long felt a greater connection with cattle and swine over birds. This may explain why her slaughter systems for cattle and pigs, unlike her system for chicken, involve unique technologies she has designed. Cattle facilities she has designed, for example, include curved handling chutes that are located between the outdoor holding pens and the slaughter facility proper (Grandin 2011). The curve in a chute encourages the cattle to walk in an arcing circle, as they do naturally in a herd, eliminating the need to strike them with cattle prods. Similarly, a cow's lack of depth perception to the side or rear means that even distant objects in those directions can appear to be within the animal's flight zone. Solid walls in the chute eliminate the possibility of the animal seeing people or other distractions that might startle them (Grandin 1980a). Another device invented by Grandin is the conveyor-restrainer. It is what is waiting for cows and pigs after the handling chute leads them into the slaughterhouse, where they are immobilized in a low-stress manner. They are then shot in the brain with a captive-bolt stunning gun, which when used effectively renders them unconscious before they are killed (Grandin 2012; 2014).

Grandin's slaughter system for chicken lacks any equivalent devices designed with bird behavior in mind. Grandin, however, stresses that the most important element of her system is not any mechanical aspect, but the way it is used. As she puts it, "the best equipment in the world is worthless unless management controls the behavior of plant employees" (2006a: 175). A key aspect of her system therefore involves plant audits. Unannounced inspectors record the success rate of various procedures throughout the animals' time inside the facility. This is this core element of her approach that Grandin has extended to facilities that hatch, raise, and kill chickens. It sees Grandin or another trained observer inspect a poultry facility and evaluate it based on observable, measurable criteria. In order to pass the audit a facility must achieve a passing numerical score on each of several critical criteria (Grandin 2009; 2010a: 161–2). The criteria that apply to poultry are:

1. Stunner Efficacy Percentage. A rate of 99% of birds rendered insensate by the stunner is acceptable.
2. Bleed machine efficacy percentage. Bleed machines kill birds by slitting their throats. A rate of 99% of birds killed by the bleed machine is acceptable.
3. Red bird cut rate. So-called red birds are dead birds whose carcasses have become discolored as a result of entering the scalder while still alive. They are killed during the scalding process, which removes their feathers. Finding a red bird without its throat cut is an indication it died in this inhumane way. One such bird death per day per facility is acceptable.
4. Broken wing percentage. For birds weighing less than 3 kg (6.8 lbs.) a rate of 99% of birds without broken wings is acceptable. For birds weighing over 3 kg. a rate of 97% with unbroken wings is acceptable.
5. Bruised thigh and drumstick percentage. An unbruised bird rate of 99% is acceptable.
6. Broken leg percentage. Zero birds can have their legs broken during the audit itself. A prior rate of 1/500 is acceptable.
7. One-legged shackling rate. Zero birds can be shackled on only one leg during the audit itself. A prior rate of 1/500 is acceptable (Grandin 2009).

Additional factors the audit takes into account include the repair of cages and whether they allow the animals enough room to all lie down without stacking on each other. Separate audits are administered for chicken farms and hatcheries. Factors audited at farms include such measures as a gait score – farms fail if less than 95% of their flock can walk without limping for ten steps – and whether birds are given enough room so that they can all move one meter away when a person walks through the confinement facility. The criteria for hatchery audits include measures such as whether the machinery that handles live chicks ever throw them onto the floor or into the trash, two outcomes that result in automatic audit failure.

In North America, most chickens are slaughtered using a process known as electric immobilization. Upon arriving at a slaughter facility birds have their feet placed in shackles and are hung upside down on a shackle line. Some chickens respond by thrashing and trying to escape as the conveyor carries them upside down and submerges their heads in a water bath. An electric current is then run through the bath to render the birds unconscious before they are removed from the water and their throats are slit. While the live-hang method is meant to ensure that all birds are unconscious before being killed, Grandin admits that this is often not the case: "that's the way it's supposed to work, but you need *very* good handling" (Grandin and Johnson 2010: 213). Even for birds that are rendered unconscious before slaughter, the process is often stressful. In part this is a result of the relaxed regulatory framework surrounding American chicken slaughter. Although the Humane Methods of Slaughter Act stipulates that before an animal can be shackled and killed it must be unconscious, the U.S. Department of Agriculture has long deemed the legislation inapplicable to the slaughter of chicken (Humane Society 2008).

An alternative method of slaughter involves gas stunning. Rather than be hung upside down, birds never leave their transport crates. The crates are conveyed through an atmosphere-controlled room in which oxygen is slowly replaced with a non-lethal gas that knocks the birds unconscious. Once unconscious the birds die due to a lack of oxygen. Grandin has long emphasized the superiority of controlled-atmosphere over electric killing (Grandin and Johnson 2010: 213). Her reasons include the fact that controlled-atmosphere killing involves less stressful handling. It also significantly reduces the possibility of birds being killed while still conscious. Controlled-atmosphere systems, however, are expensive, and to date they remain the exception rather than the rule in North America and elsewhere. As a result, Grandin's humane slaughter system is designed to be used with *either* gas stunning or electric immobilization. This distinguishes Grandin's poultry system from those she has designed for cows and pigs, which permit no equivalent flexibility regarding handling or stunning.

Grandin's method seeks to minimize animal suffering while maximizing profit. This is most evident as it pertains to cattle slaughter. Cows that go through her system have been measured to have the same low level of the stress hormone cortisol as they do when undergoing vaccination (Grandin 1998). They also receive far fewer bruises than at traditional slaughter plants (Grandin 2000). These and other factors increase the value of the animal's carcass, with the result that a humane cattle system is also a more profitable one, a consideration that Grandin has often stressed (Grandin 1980b, 2000, 2013).

According to Grandin, there are no equivalent economic factors that weigh in favor of improving poultry slaughter. "There's not very much built-in economic incentive for managers to take good care of their birds. To some degree, it's the opposite" (Grandin and Johnson 2010: 225). Grandin gives the example of stacking multiple layer chickens into cages that severely restrict their natural functions. Although the productivity of each individual bird decreases, the ruthlessly efficient use of space enables the use of more total chickens and so results in more eggs.

Singer Reconstructed

Grandin herself has offered numerous justifications for her system. Her justifications however are offered within a speciesist, and so non-protectionist framework, and so fall outside the scope of my concern. A rationale for her system, however, has been drawn from Peter Singer's work. At first glance, Singer's work would not seem a promising source of justification for killing animals. As the philosophical founder of the animal protection movement, Singer may have done more to turn people away from meat-eating than anyone alive. Nevertheless, Gary Francione and other critics of Singer argue that Singer's philosophy does not entail eschewing meat. It rather entails the widespread adoption of a system such as Grandin's, which seeks to minimize food animal suffering while permitting continued killing. If this view is correct, it means that Singer's influential argument has been widely misunderstood. While Singer himself has repeatedly called for the widespread adoption of vegetarianism, and describes his own diet as quasi-vegan, on this interpretation of his argument, neither diet really follows from his argument (Clyne 2006).

In order to understand this view, some exegesis of Singer's argument is required. That argument appeals to the equal consideration of interests. Singer demonstrates the force of this familiar principle by means of a comparison Grandin also makes, that between a mentally handicapped human and an animal with a similar or greater level of cognitive ability (Singer 1990: 19). If it would be wrong to attack the handicapped human and cause him pain, we should give the same moral weight to the interest of an animal with similar cognitive abilities not to experience an equivalent amount of pain, hence the traditional understanding of Singer's work as an argument for vegetarianism. Intensive farming systems have traditionally operated without regard to the animals' needs. Painful procedures such as beak cutting, castration without anesthetic, and immobilizing confinement have all long been standard practices. Given the pain caused by meat production, Singer has argued, vegetarianism is morally obligatory.

Francione, however, draws attention to passages in Singer's work that conditionally condone killing animals for food. Conditions include that the killing is done painlessly, and that the animal in question be replaced by another animal. As Singer puts it in the second edition of *Animal Liberation*:

> [T]he argument from the benefit of bringing a being into existence could justify continuing to eat free-range animals (of a species incapable of having desires for the future) who have a pleasant existence in a social group suited to their behavioral needs, and are then killed quickly and without pain. I can respect conscientious people who take care to eat only meat that comes from such animals (Singer 1990 230; cf. Francione 1996: 50).

Elsewhere in his writings Singer draws a distinction between killing a person and killing a merely sentient being. Singer's emphasis on this distinction is evident in his discussion of an example taken from the work of primatologist Jane Goodall. It concerns a chimpanzee named Figan, whom Goodall observed using forward planning in the wild (Singer 1993: 116; 2011: 99). This occurred when Figan saw a banana in a tree above a more dominant chimpanzee named Goliath. After noticing the banana, Figan moved away from Goliath and sat down facing away from the banana. (Goodall suggests this was to stop himself from giving away the fruit's location: chimpanzees are careful observers of each other's eye movements.) When Goliath left 15 minutes later, Figan promptly climbed the tree and helped himself to lunch. According to Singer, Figan's actions showed him to be able to form and act upon future-oriented plans. Such an ability presupposes that that animal has some conception of itself existing beyond the present. When an animal falls into that category, Singer argues, it has an interest not merely in avoiding suffering, but in not being killed. For an entity that has a sense of its own future, even a painless death would prevent the realization of its future-oriented goals or desires. An animal such as a chimpanzee therefore has an interest in continued existence. Singer argues that we should take that interest into account when deliberating over whether or not to kill the animal in question. If we were in a situation where we were deciding between a chimpanzee's life and our own, then the principle of self-defense would justify killing it. But to kill a future-conscious animal for sport or pleasure cannot be justified on the equal consideration approach, as it involves depriving the animal of its future. A person, whether human or animal, is able to form preference about the future. To kill such an individual is to thwart such preferences, which Singer argues is morally wrong (1993: 95; 2011: 80).

Creatures that are merely sentient, by contrast, do not have the ability to form preferences regarding the future, and so different considerations apply. Singer argues that in such instances the wrongness of painless killing derive from the loss of pleasure involved (1993: 132). The concern with pleasure however means that when the death of one animal makes possible the birth of another, no wrong actually occurs. This is because the loss of pleasure on the part of one animal is made up for by the pleasure of another. This is of course Singer's well-known replaceability argument. He has outlined its ramification for humane slaughter using the example of chickens:

> Let us make the questionable assumption that chickens are not self-conscious. Assume also that the birds can be killed painlessly, and the survivors do not appear to be affected by the death of one of their numbers. Assume, finally, that for economic reasons we could not rear the birds if we did not eat them. Then the replaceability argument appears to justify killing the birds, because depriving them of the pleasures of their existence can be offset against the pleasures of chickens who do not yet exist, and will exist only if existing chickens are killed (1993: 133).

It is this strand in Singer's writings that Francione has in mind when he characterizes Singer's theory as treating animals as replaceable resources (Francione 2008: 55). What makes the birds replaceable on this view is a utilitarian commitment to maximizing animal utility, understood here in preference-utilitarian terms. Insofar as one bird's satisfied preferences are replaced by another's, no moral wrong has occurred.

Francione's reconstruction of Singer's view as a rationale for humane slaughter is evident in his general summary of Singer's account: "we should focus more on animal suffering and less (and perhaps not at all) on animal killing except to the degree that it causes suffering" (2008: 18). By that criterion, Francione argues, Grandin's system of slaughter "very neatly fits Singer's exhortation that animal advocates support any measure that 'reduces the suffering of animals or enables them to meet their needs more fully'" (Francione 1996: 200). According to Francione, there are no grounds in Singer's philosophy to oppose humane slaughter in the name of veganism.

In this way, Francione articulates a view of Singer's work that many other critics have expressed. Jocelyne Porcher, for example, draws an affinity between Singer the philosopher and Grandin the slaughterhouse designer:

> In theory, Singer proposes "liberating animals." In practice, he provides McDonalds with (perhaps profitable) assistance in improving the fate of animals in industrial systems. This is also done, by the way, by the ethologist Temple Grandin, who also claims to work for animals by improving conditions in industrial slaughter (2017: 93).

J. Baird Callicott also connects Singer and Grandin in a discussion that draws attention to the fact that Singer's animal ethic is utilitarian (2016: 48). He suggests that such an ethic does not entail avoiding meat, but rather the implementation of "humane methods of transportation and slaughter," citing Grandin's guidelines as an example of such methods (Callicott 2016: 48; cf. Chambers and Grandin 2004). As Callicott writes, "those persuaded by utilitarian animal ethics can, as things now stand, eat meat in good conscience provided, like I, that they can assure themselves that the animals they are eating were raised in pleasant circumstances and were humanely transported and slaughtered" (2016: 48).

These critics disagree among themselves on our ethical obligations to animals. Francione is a traditional vegan. Porcher argues that we should eat only limited amounts of meat from animals raised at small-scale farms (2017: 94). Callicott endorses eating animals that were raised humanely. What they agree on is not that Grandin's system of humane slaughter is necessarily morally defensible, a notion Francione rejects, but that her system is morally defensible within the framework of Singer's philosophy.[3]

This reconstruction of Singer connects his philosophy to Grandin's system in way Singer to date has not. It is undeniable that Singer has supported Grandin's system as a superior alternative to humane slaughter. Singer, however, has generally characterized this support as a non-ideal compromise. While it would be better for people to stop eating animals altogether, insofar as that is unlikely to happen in the short term, it is best to combine idealistic advocacy for animal liberation with realistic advocacy for a humane death for the millions of animals who will inevitably be killed for food (Clyne 2006; Barkham 2006; Singer 2015). Francione's reconstruction of Singer's argument however denies that Singer's theory entails any reform more idealistic than humane slaughter. Singer's support for Grandin's system, rather than a pragmatic compromise, should rather be viewed as the ideal outcome of Singer's theory.

Utilitarianism vs. Equal Consideration

Francione characterizes Singer's theory as utilitarian (1996: 47–62; 2000: 135–48; 2008: 18, 54–55). This is true as far as it goes. Singer characterized persons as non-replaceable because of an underlying commitment to

[3] Francione offers his criticism of Singer by way of arguing that his own theory of animal protection is superior. Francione's positive views are however outside the scope of my discussion. For critiques of Francione, see Sunstein (2001) and Wayne (2013).

preference utilitarianism: snuffing out their future-directed preferences is wrong because morality entails maximizing preference satisfaction.[4] In the case of merely sentient beings incapable of future-directed preferences, Singer's preference-utilitarian approach comes to resemble that of hedonistic utilitarianism, insofar as the sentient beings' preferences "will be of a more immediate sort. They will not have desires that project their images of their own existence into the future" (1993: 126). Given the short term nature of such beings' preferences, they will often be directed toward experiencing pleasure and avoiding pain. The replaceability argument in this way is a direct outgrowth of Singer's preference-utilitarian framework.

What is often overlooked, however, is that Singer has also relied on a non-utilitarian argument. Singer's non-utilitarian argument is on clearest display in the first edition of *Animal Liberation*. As Singer has written, "the text of [the first edition of] *Animal Liberation* is not utilitarian. It was specifically intended to appeal to readers who were concerned about equality, or justice, or fairness, irrespective of the precise nature of their commitment" (Singer 1999: 283) There, Singer makes an argument that appeals to an ecumenical principle of equal consideration, as I did in Chapter 1. This principle as we have seen is one endorsed by a variety of moral theories. Such a principle is what Will Kymlicka draws attention to when he writes of utilitarianism and deontology that "both work from a basic moral commitment to the idea of impartiality. Both theories accept that, from the moral point of view, each person is equally worthy of moral consideration, each person is an end in herself, whose interests must be given equal consideration" (1991: 159). The reason non-utilitarians can accept Singer's argument in *Animal Liberation* is because it extends the ecumenical equality principle to apply beyond persons to merely sentient beings. As Singer notes there, what the principle will entail for our treatment of different animals will vary according to their capabilities and the interests those capabilities give rise to. "But the basic element – the taking into account of the interests of the being, whatever those interests may be – must, according to the principle of equality, be extended to all beings, black or white, masculine or feminine, human or nonhuman (Singer 1975: 5).

Call Singer's non-utilitarian argument the equal consideration argument. Although Francione does not appeal to it, it bears noting how it

[4] Singer has more recently rejected preference utilitarianism in favor of hedonistic utilitarianism as the right theory to apply to merely sentient beings and persons alike. See further discussion, following.

too might be invoked to justify humane slaughter. The primary rationale for doing so is that the equal consideration argument employs fewer controversial assumptions than utilitarianism, a feature that goes some way toward establishing its status as the stronger of Singer's two arguments. In particular, the equal consideration argument does not employ the replaceability argument, which as we will see has highly counterintuitive implications. Singer himself has sometimes blurred together the two arguments. I have elsewhere taken up the scholarly task of exegetical separating the two arguments (Lamey 2011). Here, however, I focus entirely on the philosophical project of deriving the strongest possible rationale for humane slaughter from Singer's texts. In doing so I consciously go beyond the reconstruction Francione has outlined and offer what I take to be a philosophically stronger reconstruction of my own.

Singer's non-utilitarian argument appeals to two notions also employed by his utilitarian argument. The first is the equal consideration of interests. This view of equality should be familiar, as it was the one I employed in my opening chapter. As Singer summarizes this view of equality, it amounts to the belief that "the interests of every being affected by an action are to be taken into account and given the same weight as the like interests of any other being" (Singer 1990: 5). As we saw, the equal consideration of interests is endorsed by a wide variety of moral and political philosophies. Although they include utilitarianism, they also include deontological theories and other traditional rivals to utilitarianism (Kymlicka 1991: 159, DeGrazia 1999: 112). Different theories disagree as to how equal consideration should be applied or whether morality needs to take into account other principles as well. One possibility, for example, is the equal consideration governs our political deliberations, but we need to give some weight to partiality in some areas of personal morality (Nagel 1991). Very few moral theories grant no weight whatsoever to equal consideration, however, and it is rightfully regarded as an ecumenical rather than exclusively utilitarian principle.

The second notion that is found in both Singer's utilitarian and non-utilitarian arguments is the distinction between two levels of moral standing, one pertaining to persons the other to merely sentient beings. Just as many moral theories other than utilitarianism make appeal to equal consideration, so too have non-utilitarian philosophers given weight to whether or not an entity can conceive of itself over time. Many definitions of moral personhood for example, give weight to such a consideration. Such is the case with Michael Tooley's widely discussed account of personhood, which holds that only moral persons have a right to life. As Tooley summarizes his view:

> An organism possesses a serious right to life only if it possesses the concept of a self as a continuing subject of experiences and other mental states, and believes that it is itself such a continuing entity (Tooley 1972: 44).

An ability to conceive of one's self across time is also regarded as significant by J. David Velleman. As we saw in Chapter 2, Velleman argues that because a cow cannot possess such a conception, it cannot care about its future self. In this way, Velleman's theory emphasizes an ability to conceive of one's life in its entirety, an aspect not present in Tooley's account. Tooley, for his part, speaks of a right to life, a notion that Singer in his utilitarian mode consciously rejects. In spite of these and other important differences among their various theories, Singer, Tooley, and Velleman all ascribe philosophical significance to an entity's ability to conceive of itself as persisting through time, as do many other philosophers (e.g., McMahan 2002: 194–6).

As with the equal consideration of interests, it is possible to debate the ethical significance of self-conception through time. One reasonable question to ask for example is whether it is a necessary condition of moral personhood, as Singer, Tooley and others have argued, or a sufficient one.[5] Few ethicists, however, reject it as an entirely irrelevant consideration, and it seems plausible to hold that, whether or not we are utilitarians, we have a pro tanto moral obligation not to kill an entity that can form expectations about the future, regardless of whether or not we also have a similar obligation to sentient creatures without a sense of their own future.

So, Singer's equal-consideration argument shares with its utilitarian counterpart a commitment to the equal consideration of interests and a two-level theory of moral standing that distinguishes persons from merely sentient beings. One aspect of Singer's utilitarian argument that the equal-consideration account does *not* share, however, is the replaceability argument. That argument, as we saw, was a direct outgrowth of preference utilitarianism. Appealing to it would defeat the purpose of the equal-consideration argument, which is meant to have force against deontologists, proponents of common sense morality, and other non-utilitarians.

[5] Such a view would head off an endorsement of infanticide, which Tooley and Singer both argue does no wrong to the infant (Tooley 1983: 412, Singer 1993: 182). See Bermúdez 1996 for an argument against infanticide that still permits abortion.

So what does the equal-consideration argument say about killing a merely sentient being? Singer offered the equal-consideration argument in the first edition of *Animal Liberation*, which was deliberately silent on the ethics of killing. As he put it, "the conclusions that are argued for in this book flow from the principle of minimizing suffering alone" (1975: 22). Singer adopted this approach in part because he did not consider it necessary to outline an ethic of killing to call into question meat eating and other practices. When the equal consideration principle was applied to animal suffering that was enough to challenge "all the major abuses of animals that human beings practice" (1990: 17).

An implication of this silence on the ethics of killing animals however is that the equal consideration argument can potentially be used to justify humane slaughter. This is because it is designed to be compatible with a range of views, utilitarian, deontological and otherwise. The permissibility of killing animals so long as it is done painlessly is affirmed by a wide range of theories that affirm a two-level theory of moral standing. Examples include the theories of Tooley and Velleman already noted, both of which make such an allowance on broadly similar grounds to Singer in his utilitarian mode. Both theories concur that because animals lack a sense of themselves existing through time they lack an interest in continued existence (Tooley 1972: 60–2; Velleman 1991: 71). As Velleman puts it, "a premature death [does not] detract from the value of [a] cow's life as a whole, since a cow has no interest in its life as a whole, being unable to care about what sort of life it lives" (1991: 71). Tooley and Velleman differ in the details, but the basic idea they share is one that is part of common sense, as evident in the fact that Grandin also assumes that animals are not harmed by death. Insofar as the equal-consideration argument is deliberately designed to be compatible with views that permit the painless killing of merely sentient animals, then it would seem to permit painless killing as part of human slaughter.

When Singer first published *Animal Liberation*, slaughter was far from humane. Perhaps this is one reason Singer did not think it was necessary to appeal to the interest animals have in continued existence, as opposed to their interest in avoiding suffering. For years afterwards Singer took a concern with animal suffering as sufficient grounds to prohibit omnivorism (1990: 230; 1993: 134). The arrival of Grandin's system, however, introduces a potential complication into this picture. It raises the possibility that painless killing, if it is not already the rule, will someday become the rule rather than the exception. Such a change raises the possibility that Singer's equal consideration argument permits rather than prohibits eating animals, as mainstream animal agriculture will eventually meet the conditions the equal-consideration outlines for killing animals for food and other purposes.

Singer's utilitarian argument will of course be of interest to utilitarians. His equal consideration argument will be of interest to many non-utilitarians. If the revised view of Singer's theory is correct, one of the most historically powerful arguments for vegetarianism will be shown in the end to legitimize a form of meat consumption that is increasingly widespread. This is an outcome ethical vegans and animal advocates cannot help but care about, whether or not they frame the case for animal protection precisely the way Singer does. The relevance for meat eaters is, if anything, even more apparent. If their diet is challenged on ethical grounds, they can reply that one of the world's leading animals ethicists says it is permissible so long as the animals are killed painlessly, a possibility which Grandin's system is widely thought to realize on a widespread scale.

Criticism

What are we to make of humane slaughter and the preceding defenses of it? As a feat of engineering, Grandin's system of slaughter is nothing short of remarkable, exhibiting as it does an inspired combination of technical ingenuity and insight into mammalian behavior. Grandin's designs are based on empathetic insight into animal perception. Whether or not one thinks the empathetic element extends far enough should not stand in the way of recognizing what a progressive step Grandin's system represents against the backdrop of traditional agriculture.

A full assessment of Grandin's system, however, requires taking note of its gaps and limitations. To that end, one aspect that that bears pointing out is that Grandin's system aims for a lower frequency of painless killing than is formally required by U.S. law. The 1958 Humane Methods of Slaughter Act legally requires that all pigs and cows killed for food be unconscious at the time of death. That the law was never enforced explains how Grandin's system could represent an improvement over what came before (Jones 2008). Grandin however maintains that 100 percent painless kill rate is not possible. As a 2004 government report Grandin was involved with put it, "Dr. Grandin believes that effectively stunning animals on the first try 100 percent of the time is unachievable – that is why she proposed an objective scoring method as an alternative" (Government Accountability Office 2004: 18; Grandin 2006b: 133).

This is an important point that is often overlooked in debate around humane slaughter. Painless slaughter was thought for several decades to be an appropriate standard to aim for from the ideal point of view. It remains in principle, if not at the level of enforcement, the standard of American law. Grandin, who provides perhaps a more detailed grasp of industrial slaughter than anyone else alive, urges that we accept the inevitability of suffering as part of animal slaughter. As much as her system seeks to reduce suffering in practice, therefore, at an ideal level it simultaneously represents a greater tolerance of animal suffering. This is because of the various forms of animal suffering in death that Grandin's system permits.

Humane slaughter's tolerance for animal suffering is particularly evident in the case of chicken. In the layer industry, male chicks have no economic value due to their inability to lay eggs. Therefore, it has long been industry practice to kill them within 72 hours of birth. Standard methods for doing so include sending the live chicks through a macerator, which is a high-speed grinder. In 2000 the number of male chicks killed this way in the United States every year was estimated at 260 million (Metheringham 2000: 25, 27). Technology has subsequently been developed that can identify the sex of the chick while still in the egg and so allows their eggs to be disposed of before hatching (Weissmann et al. 2013). Maceration, however, is less expensive and remains common. Grandin's hatchery guidelines permit the use of both controlled atmosphere killing and maceration. As such, they do nothing to address the painful slaughter of hundreds of millions of male chicks, who are ground to death while fully conscious.

The enormous numbers of chickens involved in the poultry industry means that Grandin's system permits significant suffering in other ways. Farms can pass if five percent of birds cannot walk evenly for ten steps, or if three percent of birds have broken wings. This means that on a national level, the system is compatible with millions of birds suffering from lame or broken limbs. Similarly, Grandin identifies as a "major welfare problem" the fact that chickens, turkeys, and chicks often have various parts of their bodies amputated, including their beak, toes, and combs, in processes that do not involve anesthetics or painkillers (Grandin and Johnson 2010: 212). Grandin notes that debeaking has in recent years been made more humane by the development of new technology that removes the beak with an infrared beam (Grandin and Johnson 2010: 215). Even the revised practice, however, is traumatic for the animal and still results in the loss of an important sense organ.

Chickens are debeaked to prevent feather pecking and cannibalism, practices that Grandin says are exacerbated by industry breeding practices that over-select for size alone. Her system addresses feather pecking and cannibalism by scoring the percentage of birds that have good feather conditions, but it does little to address the underlying breeding practices that lead to these conditions. As for the ways chicken's bodies are mutilated other than debeaking, Grandin's system operates on the assumption that they are standard industry practice. As she writes, "the problem of doing painful, invasive procedures on chickens without anesthetics or painkillers isn't close to having an answer" (Grandin and Johnson 2010: 214). And in the moment of slaughter itself, Grandin's similarly assumes that a small percentage of chickens and other animals will die in pain. As one NGO report pointed out, "[e]ven if 100 percent of slaughter plants were able to meet [Grandin's] standards, it would meant that 185 million chickens, 1.8 million cattle and sheep and one million pigs may be killed inhumanely each year in the United States" (Jones 2008: 86). Grandin's method ultimately confirms something critics of industrialized animal killing have long maintained. Suffering is an inescapable part of the process.[6]

All of the aforementioned types of suffering will occur even in scenarios in which Grandin's standards are rigorously enforced. But because her standards are enforced through audits, there are built-in limits to how rigorously they can be enforced. For example, a perennial issue in the slaughter industry has been the speed with which animals are killed. Whereas cattle slaughter facilities killed 50 cows an hour thirty years ago, recent decades have seen some plants routinely kill over 400 animals an hour (Warrick 2001). The economics of the slaughterhouse industry provide incentives for operators to kill as many animals in as little time as possible. Grandin has pointed to the phenomenon of plants processing animals at a greater speed than their facilities are designed to handle as one of the scenarios most likely to give rise to animal abuse (Grandin 2013: 8). An occurrence her system cannot entirely prevent is that of plant managers deliberately slowing down their line speed when auditors arrive. Although Grandin's audits are unannounced, inspectors must still typically present themselves at the facility's gate upon arrival, put on safety equipment and take other steps that delay their entry onto the kill floor. Critics contend

[6] Grandin's system also currently does not involve any labeling program. This means that unless one eats only meat from McDonald's or Burger King, which seems unlikely, there is no way for consumers to know when they are buying meat from animals killed in one of Grandin's facilities.

that this gives plant managers enough time to slow down the line. (Eisnitz 2006: 1; Washington State Department of Agriculture 2001: 3)

Grandin has not disputed that this takes place. "By the time I get there of course they're on their best behavior," she has said (Case 2010: 1). During an investigation into a Washington state plant that was accused of committing gross abuses even though it had passed Grandin's audit, Grandin told a reporter that she was aware at the time of her audit that the slaughterhouse in question had deliberately slowed down its line speed during her visit (Lynch 2001: A1).[7] Grandin's auditing reports, however, do not address this issue, and are instead presented as if what the auditor sees is typical of what happens when auditors are not around. In cases in which that condition is not met, the gap between traditional slaughter and humane slaughter narrows.

As we saw, Grandin's approach is most well established in the case of cows and pigs. Partly this is because of the economic structure of the meat industry, in which there is a greater diversity of beef and pork supplies than of chicken suppliers. As Grandin explains, "one of the reasons why McDonald's and other hamburger chains were effective in improving the U.S. beef slaughter plants is they had more than 40 plants on their approved supplier list. If one or two plants failed an audit, they could stop buying from them and still have enough beef" (Grandin 2013: 5). Grandin notes that less progress has occurred with chicken plants, because it is common for restaurant chains to have fewer chicken suppliers than in the case of beef and pork. This limits the clout that chicken buyers have. As for fish or exotic meats, Grandin's methods do not extend to them. Grandin's system is thus most effective in the case of pork and beef, two meats that have declined as portion of the average North American diet in recent decades, and which now represent less than 5 percent of food animals killed.

Food animals can live for years but typically only spend a few hours at a slaughter facility. Grandin's system does not address forms of suffering that can take place prior to slaughter. These include practices such as castration, branding, animal fighting, and intensive confinement. Grandin's guidelines also say nothing about what an animal is fed prior to slaughter or issues having to with the manipulation of an animal's size and body structure. Grandin notes that it is now common for dairy cows to be bred at such a size their feet can no longer support their body weight

[7] The plant in question was ultimately cleared of the violations in question. See Washington State Department of Agriculture (2001).

(Grandin 2001: 107). Grandin's approach has to date been most effective in implementing reforms that enhance profitability, such as reducing bruising. It is less likely to lead to welfare reforms that require economic sacrifice, which are likely to require legislation.

Finally, failing a Grandin audit is not a very consequential event. As Grandin notes, "Most plants that fail an audit are not permanently 'delisted.' After they make improvements and pass a re-audit they are put back on the approved supplier list." (Grandin 2010b: 219). The system does not challenge the widespread notion that arrangements that would be unthinkable for human beings are appropriate in the case of sentient animals. It is perhaps little wonder that voluntary programs such as Grandin's "are commonly cited by agribusiness during legislative deliberations and used to argue that it is not necessary to pass legislation to prevent cruel farming practices" (Farm Sanctuary 2005: 3). They require little sacrifice on the part of the industry, and involve temporary consequences when they are violated.

The Singer Rationales Re-examined

We are now in a position to assess the claim that Singer's work provides a rationale for humane slaughter. To provide a rationale in the relevant sense is to affirm that humane slaughter lives up to the standards of permissible animal killing as outlined either by Singer's utilitarian argument or his equal-consideration argument. It should be obvious that Grandin's system fails to meet the standard of permissibility advanced by either argument. In both cases animal slaughter was impermissible when accompanied by suffering. As we have just documented, humane slaughter permits significant amounts of animal suffering, including during slaughter; does not address suffering that occurs before slaughter; is most effective in cases involving less than five percent of food animals, and even then provides an uncertain picture of what takes place at the time of slaughter; and involves limited consequences for plants that fail to meet Grandin's guidelines. Taken as a whole, these considerations suggest that Grandin's system fails to find a rationale in either argument Singer has offered. We should reject Francione's suggestion that Singer's work contains no grounds on which to criticize humane slaughter.

But what about a future version of humane slaughter? The rise of Grandin's system is a reminder of the historically changing nature of the slaughter industry. It is at least conceivable that a system may someday arise that involves the slaughter of enormous numbers of animals done painlessly without exception. So let us imagine a future version of a humane system that eliminates all suffering for every animal. Would either Singer's utilitarian argument or his equal consideration argument justify that form of humane slaughter?

To answer this question we need to recall a distinction that both of Singer's arguments employed. This was the distinction between persons and merely sentient beings. According to both arguments, it was a condition of painless killing being permissible that the creatures in question have no awareness of their own future. This is a conceptual criterion. A second aspect of Singer's analysis, however, concerns which particular animal species lack a sense of their own future. This is an empirical criterion. Singer's understanding of which animals meet it has changed over the years, in keeping with new findings regarding the cognitive abilities of animals.

Singer has long suggested chimpanzees, gorillas, and orangutans represent the clearest cases of animals which possess future-oriented desires (Singer 1993: 118; 2011: 101).[8] More recently, Singer has suggested that it is not far-fetched to ascribe such desires to a species of octopus, and that magpies and parrots are capable of a broadly similar capacity for self-awareness (2011: 103). In the case of whales, dolphins, monkeys, dogs, cats, pigs, seals, bears, cattle, and sheep, he concludes that the evidence is ambiguous, but we should give the animals the benefit of the doubt (Singer 1993: 118–9; 2011: 103). In regards to chickens, as we saw, Singer has long mooted the possibility that they do not have a sense of themselves existing through time. Singer more recently has recently taken note of studies that ascribe to chickens rudimentary aspects of self-consciousness (which we will discuss in the next chapter) (Singer 2011: 120). Even after noting such studies however Singer still leaves open the possibility that chickens can be killed for consumption. "Suppose we could be confident that chickens, for example, are not aware of themselves as existing over time (and as we have seen, this assumption is questionable). Assume also that that the birds can be killed painlessly, and the survivors do not appear to be affected by the death of one of their numbers" (2011: 120). If all of those conditions hold, Singer concludes, then his replaceability argument still "appears to justify killing the birds" (2011: 120).

[8] In the first edition of *Practical Ethics,* Singer ranked apes, whales and dolphins as the animals with the highest level of cognition (1979: 103). The change in status for whales and dolphins is presumably a result of Singer re-evaluating the empirical evidence in their cases.

This view of animal consciousness is more nuanced and qualified than Singer is sometimes given credit for. Francione, for example, suggests that Singer views animals in general as merely sentient. According to Francione, "Singer regards most nonhumans as living in a sort of eternal present that precludes their having an interest in a continued existence ... [Singer assumes] that animals do not have an interest in their continued existence" (2008: 18). In fact, Singer's view precludes general declarations about "most nonhumans" or animals as such. Everything comes down to what particular species we have in mind, and what evidence is invoked to show that the species in question either lacks or possesses a sense of its own future.

Singer notes that in the case of some species, the evidence for their self-consciousness can be ambiguous. Singer specifies that when there is inconclusive evidence as to whether an animal has a sense of its own future, the moral thing to do is to give it the benefit of the doubt, and treat it as if it does (1993: 119). This aspect of Singer's argument appeals to a notion we frequently employ in our day to day moral reasoning, a concern not merely with harm, but the risk of causing harm. It is a reasonable proviso for Singer to attach to his criterion of painless killing. Even when the evidence in a given case is suggestive rather than conclusive, it provides grounds on which to give the species in question the benefit of the doubt, and so entails not killing them.

Singer has long suggested that we give the benefit of the doubt to pigs, cattle and sheep (1993: 132). More recently he has cited contemporary findings regarding the cognitive abilities of birds and even fish to suggest that there may be a weaker case for extending the benefit to them as well (2011: 120). In order for either of Singer's arguments to be used to justify humane slaughter, even under the idealized circumstances we are now considering, much will depend on how the benefit of the doubt standard is applied. Much will also depend on which particular species of food animal is being proposed as a candidate for humane slaughter. These considerations do not rule out the very possibility that idealized humane slaughter may prove justified under either of Singer's arguments. Nevertheless, they do place obstacles in the way of such a case. In particular, they oblige a proponent of the reconstructed Singer argument to provide reasonable evidence to the effect that a given species of food animal lacks any sense of its own future. Even Francione, the leading proponent of the reconstructed view of Singer, has not gone this far. As such, the claim that Singer's view justifies humane slaughter in general, as it pertains to all the familiar species of food animals, remains open to doubt. It is not so much refuted as incomplete, as two important burdens of proof, concerning the appropriate application of the benefit of the doubt standard and the empirical evidence to indicate a given species is not harmed by death, have not been met.

This still leaves open the possibility that some particular species of food animal will prove to be merely sentient. Even if all the species previously mentioned displayed some degree of self-consciousness, that would still leave the possibility that crustaceans, for example, were merely sentient. Alternatively, we might imagine that fish or chicken or other animals mentioned by Singer prove upon further investigation to be merely sentient and lack any trace of self-consciousness. So let us now finally ask what Singer's arguments would say in this circumstance of double idealization, under which we imagine a system of humane slaughter that truly is painless, and the merely sentient status of the species in question is not in doubt.

Under the conditions of double idealization, both Singer's argument would justify humane killing. They differ in the conditions that would have to be met and in their implications. Regarding Singer's utilitarian argument, humane killing would be justified if other animals were brought into existence to replace those that are killed. Under the equal consideration argument, humane slaughter would be justified whether or not the animals in question were replaced. Thus if we imagine someone killing an animal in Robinson Crusoe-like circumstances, where resources are limited and they do not have the ability to replace an animal that they kill, it would be easier to justify on the equal consideration argument than the utilitarian argument. In normal conditions where replacement is possible, the equal consideration argument imposes a less demanding standard, one that leaves it up to the slaughterer whether replacement animals should be born. The equal consideration argument is thus the more permissive of the two. It is also the more difficult of the two arguments to reject, insofar as it will be unscathed by criticisms that zero in on the argument or the hedonistic utilitarianism on which replaceability rests. Both arguments however ultimately fail to justify humane slaughter, for reasons I will now try to show.

Let us begin with the utilitarian argument. Singer invokes it to argue that parents of a severely disabled infant can kill it if after doing so they replace it with a second child who will have a more pleasant existence (1993: 181–91; 2011: 160–7). This is only consistent: if the replaceability argument justifies killing one animal non-person and, all else being equal, replacing it with another that will live an equally pleasant or more pleasant life, given anti-speciesism, it will also justify doing the same in the case of human non-persons, all else being equal. This aspect of Singer's argument has long attracted controversy (Singer 1993: 337–59). What has not often been noted, however, is that the argument actually permits the replacement of any infant, disabled or not, by a second child who will have a more pleasant existence. As Singer notes, the replaceability argument recognizes no difference in the moral status of disabled and other infants. "The difference between killing disabled and normal infants lies not in any supposed right to life that the latter has and the former lacks, but in other considerations about killing. Most obviously, there is the difference that often exists in the attitudes of the parents" (1993: 182; 2011: 161). In cases where the birth of a non-disabled child are a "threat to the happiness of the parents," all the considerations in favor of permitting killing and replacing a severely disabled infant also permit killing and replacing a non-disabled infant.

I take it for granted that this is an unattractive feature of the utilitarian argument. There is, however, a second problem with the argument. It concerns the application of utilitarianism's maximizing logic to questions involving conception. One way to maximize utility is to make existing people happier. Another way is to create more happy people. The replaceability argument is based on a utilitarian commitment to maximizing utility through the creation of happy lives. This suggests that we can be morally criticized for failing to maximize the number of happy lives in existence. Intuitively, this is hard to accept. We do not normally think it is a strike against vows of chastity or the use of birth control for example, that such practices could fail to maximize the number of people in existence. To be sure, questions involving conception, which we will analyze at length in Chapter 7, are some of the most difficult to answer. Yet as complex as the underlying issues are, I cannot help thinking that this particular implication, which cuts deeply against our normal approach to questions related to conception, is especially worth avoiding.

A third and final concern raised by the replaceability argument is the scope of its application. As Singer originally presented the argument, it did not entail the replaceability of persons. It is open to question however whether Singer really can restrict the scope of the argument in this way. Singer sought to avoid this outcome by arguing in the case of persons that utility should be thought of as preference satisfaction. Insofar as persons can form preferences regarding the future, death is wrong because it frustrates such desires. But why should this distinguish persons from merely sentient beings (Hart 1980)? Utilitarianism takes a maximizing approach to utility, whether utility is conceived of as pleasure or preference-satisfaction. So why not think that it is also permissible to replace one person with another whose preferences have an equal or better likelihood of being satisfied?

Singer has answered this question by positing a difference between satisfying someone's preferences and bringing into existence a different individual. When someone has a preference that is not satisfied, that is clearly negative according to preference utilitarianism. That negative outcome might be made up for if creating a new individual resulted in the straightforward creation of a satisfied preference. But this is not quite what happens. Rather, we create a new preference that can subsequently be satisfied. Singer has suggested that we think of new desires as analogous to debits in a ledger (Singer 1993: 128). They can be made up for, or cancelled, by being satisfied. But the overall outcome of creating a new preference that is later satisfied leaves the world neither better nor worse than it was before the preference was created. Thus whereas frustrating a pre-existing preference is bad, creating a new preference that is satisfied is morally neutral.

This strategy for avoiding the replaceability of persons, however, raises a separate problem. It suggests that it is morally wrong to bring into existence a child who, however happy their life may be overall, has at least some unsatisfied preferences (Singer 1993: 129). As we all have at least one unsatisfied desire in the course of our life, the argument entails that it would be best if none of us had been born. Singer modifies his account so as to avoid this implication:

> Thus the moral ledger model of creating and satisfying a preference will not do. It might be saved by attaching to it a stipulation that sets a given level of preference satisfaction, below complete satisfaction, as a minimum for overcoming the negative [ledger] entry opened by the creation of a being with unsatisfied preferences. This might be the level at which we consider a life worth living, form the perspective of the person leading that life. Such a solution seems a little ad hoc, but it may be possible to incorporate it into a plausible version of preference utilitarianism (1993: 129)

On this modified version of the moral ledger account of preference satisfaction, a life with some unsatisfied preferences can still be worth living. We can debate the exact amount of unsatisfied preferences that are tolerable. But wherever we place the threshold of acceptable frustrated desires, there is some point above which a life is worth living, and by extension worth starting.

The problem with this modification to Singer's utilitarian argument is that it threatens to make persons replaceable (Višak 2013: 62). The reason persons were said to not be replaceable was that creating a new preference that could later be satisfied was morally neutral. This made it impossible for them to cancel out the moral wrong of frustrating an already existing desire. This neutral view of desires that we bring into being and subsequently are satisfied suggested that no amount of them could make up for the frustration of one existing desire – they simply can't be quantified or weighted in the same terms as existing desires that are frustrated. The introduction of the idea of a "level" of frustrated desires that is tolerable, however, suggests that the overall amount of satisfied desires a life contains is relevant to determining whether such a life is worth creating. Indeed, it is difficult to avoid the thought that the fewer frustrated preferences a life contains, the better it is, according to Singer's threshold argument. But if lives are quantifiable in this way, then it seems to follow that a life that is overall *better* in terms of preference satisfaction can make up for a life with fewer satisfied preferences.

In other words, Singer's threshold argument holds that the overall amount of unsatisfied preferences a life contains is what determines whether it is or is not worth starting. The same standard of comparison however opens the door to thinking that one life above the threshold can be ranked better or worse than another life that is also above the threshold but contains a different ratio of satisfied and unsatisfied preferences. And if the ratios of satisfied and unsatisfied preferences in potential people's lives are comparable in this way, there is no barrier after all to thinking that a potential person with more satisfied preferences can replace an existing person with fewer such preferences.[9]

Since formulating the version of the replaceability argument I have analyzed, Singer has outlined several modifications to the argument meant to address its various problems.[10] Singer's most recent view has involved rejecting preference utilitarianism in favor of hedonistic utilitarianism. This shift marks an important change in the replaceability

[9] The argument in this paragraph is a modified version of one made by Višak (2013: 62–3). Višak argues that Singer's threshold argument presents him with a dilemma: "either a being is replacable, or it would be better off having not lived" (62). Višak suggests that under the replaceability argument only entities that have lives of positive value, e.g., more pleasure than pain, are replaceable. My own view is that the replaceability argument entails a *pro tanto* obligation to replace such beings with happy (or at least happier) counterparts. Thus, unlike Višak, I do not see two horns of a dilemma so much as one unwanted outcome. Višak's view derives from her antecedent claim that the replaceability argument should be distinguished from the Logic of the Larder argument, a version of which I discuss in Chapter 7.

[10] In the Third Edition of *Practical Ethics*, Singer presents a new theory of moral value, one that combines value rooted in the satisfaction of preferences with value that is independent of preferences. For cogent criticisms of this view, see Višak (2013: 64–6).

argument, insofar as it no longer cites the future-directed preferences of persons as grounds to regard them as irreplaceable. "If we accept hedonistic utilitarianism rather than preference utilitarianism," Singer has recently written, "the distinction between persons and other conscious beings becomes less significant. Instead, replaceability becomes acceptable, at least in theory, for humans as well as for animals" (Singer 2016: 235). By "humans" Singer here has in mind human persons. I believe Singer is right to conclude that the replaceability argument ultimately cannot be formulated so as to apply to merely sentient beings but not to persons. Rather than extend replaceability to persons, however, I take this as grounds to reject the replaceability argument and the utilitarianism from which it derives.

This leaves Singer's equal consideration argument. It differs from the time-relative interest (TRI) account I outlined in the previous chapter. On the reconstructed argument from equal consideration, animals are harmed by death only if they are persons. On the TRI account, animals are harmed by death *whether or not* they are persons. For this reason, it requires a justification for killing animals even painlessly, a justification that humane slaughter would not seem to meet. Outside of contexts in which no plant-based foods are available, our interest in eating animals is comparatively trivial, while their interest in continued existence is strong. I have said that the equal consideration argument advanced by Singer is a more powerful one than its utilitarian counterpart. Even so, I want to conclude my discussion by trying to bring out why the TRI account I have defended is stronger still.

One reason for preferring the TRI account is that it is concerned with both animal suffering and killing. Theories that hold animal suffering is morally wrong but killing is not have an odd implication when they are taken to their logical conclusion. Insofar as avoiding animal suffering is morally significant but keeping them alive is not, such theories seem to suggest that the morally appropriate outcome is to kill more or less all domesticated animals. Large numbers of wild animals would likely need to remain alive in order to ensure the continued functioning of ecosystems, at least those on which human life depends. But the correct stance of pets that are not persons, for example, would seem to be that we should kill them all to avoid the possibility of future suffering (McMahan 2002: 201).

The reconstructed equal-consideration argument has this unattractive implication because it holds that the suffering of animal non-persons matters morally but their continuing to live does not. That some pets may turn out to be persons in the philosophical sense would not save the theory. For surely the case for not killing all of the world's cats and dogs does not depend on whether or not there is conclusive evidence for their personhood. It is surely more plausible to say that their lives matter morally insofar as they are sentient. The separate question of whether or not they are persons would make killing them even worse if answered affirmatively; but it is not a plausible condition of their lives mattering in the first place.

There is an additional reason to prefer the TRI account to Singer's equal consideration account. It comes into view when we recall that the reconstructed Singer argument is anti-speciesist. An implication of accepting the reconstructed Singer view regarding humane slaughter is accepting something like Singer's replaceability argument regarding disabled infants. Like Singer's utilitarian argument, his equal consideration argument is "at odds with the virtually unchallenged assumption that the life of a newborn baby is as sacrosanct as that of an adult" (1993: 170; 2011: 151). As with animals, the reconstructed version of Singer's equal consideration argument entails killing human beings, whether infants or not, who are merely sentient. Insofar as we imagine a human being born with mental disabilities so severe that they do not possess personhood, the equal consideration argument entails that it is not merely permissible, but *prima facie* obligatory, to kill them. All of the reasons that apply in the case of animals also apply in the case of humans.

This outcome already seems bad enough. Surely we do not have a moral obligation to end the lives of these human beings, any more than we do to end the lives of animal non-persons. But the scope of the argument's implications are potentially even more sweeping. All infants, not merely those with rare disabilities, are not yet persons at the time of birth. Personhood, rather, is something we grow into in the months, perhaps years, after birth. It is therefore a serious question under the terms of the reconstructed equal-consideration argument whether or not we should kill every human infant, in order to prevent their future suffering, given that it matters but their future existence does not. The infant's potential suffering would be a *pro tanto* reason to end its life. Parents who did not see a *pro tanto* reason to kill their infants would generally be guilty of a moral shortcoming. Certainly there would be no reason to keep the infant alive for its own sake.

One possible means of avoiding this deeply alienating conclusion would be to distinguish non-disabled infants from animal non-persons on the grounds that infants are potential persons. Such potentiality might be thought to place them in a different category, such that we should grant moral significance to their future lives as persons.

Singer himself has rejected this way of distinguishing infants and animal non-persons (1993: 152–6). As Singer puts it, there is no rule that says a potential X has the same rights as an X, any more than the fact that Prince Charles is a potential king grants him the rights of a king. It is conceivable that someone might embrace the equal consideration argument and reject the argument against potentiality. Doing so has the familiar implication of extending a similar moral status to fetuses, which are potential persons just as much as infants. On this view, human organisms acquire moral standing at the moment of conception if not earlier: depending on how we conceive of potentiality, unfertilized ova and sperm are also potential persons, albeit in a more conditional way. This seems hard to accept. A fertilized egg at the moment of conception lacks sentience. On the view we are now considering, it would have a right to life that a sentient animal would not. It is a strength of the TRI account that it maintains the opposite view. Fetuses prior to the acquisition of sentience have no moral standing, but sentient humans and non-humans do have such standing.

In short, a version of the equal consideration argument that endorses the potentiality argument draws the sphere of moral standing too widely by including fertilized ova. A version of the argument that rejects the potentiality argument draws that sphere too narrowly, by excluding all infants, disabled or otherwise, from the category of beings whose lives matter for their own sake. In the latter case the permissibility of killing would be sensitive to the feelings of persons with bonds of attachment to the nonperson in question. But insofar as parents wanted to end the life of their own infant, or someone who lacked the attributes of personhood due to mental handicap, aging or disease, doing so would normally be permissible, whether or not it lead to the birth of a person who replaced the nonperson in question. It is hard not to see this counterintuitive implication as reason to reject the reconstructed Singer view. Certainly it seems too high price to pay for justifying slaughter, even if humane.

The TRI account I defend shares two fundamental principles with both Singer's utilitarian and equal consideration arguments. All three arguments embrace anti-speciesism and the equal consideration of interests. Unlike Singer's two arguments, however, mine employs the notion of a time-relative interest. A result of this is that merely sentient animals have an interest not merely in avoiding suffering but in continued existence. In this way, my account maintains that a moral wrong occurs when an animal is killed for such a trivial purpose as our wanting to eat them when plant-based alternatives are widely available. For this reason my account deems humane slaughter morally wrong.

In this way, my account is at odds with how many, if not most, people regard humane slaughter. But in other ways I believe my account is harmonious with widely accepted views. Velleman, whose view of animals I cited in reconstructing Singer's equal consideration argument, sought to justify his view by appealing to the widespread view that killing animals is permissible. "[My] conception of a lower animal's interests is supported, I think, by its fruitfulness in accounting for our intuitions about the moral difference between killing animals and killing people" (1991: 71). The shared intuition Velleman appeals to here is that of the average meat-eater. Velleman, however, overlooks the many counterintuitive outcomes of positing the permissibility, even in non-utilitarian terms, of painlessly killing merely sentient animals. I have been at pains to show that such a view, among other unwanted implications, entails an obligation to kill a large proportion of pets and other domesticated animals. Similarly, Velleman's view appears to presume speciesism. Once we reject speciesism, the question of the permissibility of killing animals, whether pets or live-stock, becomes inseparable from the permissibility of killing infants, some mentally disabled humans and other merely sentient beings. I believe it is a strength of the TRI account that it has fewer unwanted implications for ethical questions pertaining to entities other than food animals.

Conclusion

I have examined two possible justifications for humane slaughter that can be reconstructed from Singer's work and argued that neither succeeds. Humane slaughter is on my account is morally wrong. There are, however, degrees of moral wrongness. Nothing in my account denies that humane slaughter is superior to traditional slaughter. If it seems unlikely that Grandin's system has taken all suffering out of animal slaughter, it seems even more unlikely that it has made no difference, either. The handling chutes and other elements that reduce an animal's stress in its final hours are improvements over previous slaughter systems, which did nothing to reduce the terror animals experienced before death. For this reason, pointing out problems with Grandin's approach at an ideal level should not be taken to show that nothing is gained when plants adopt her approach. As much as we might hope for a world in which McDonald's sold only vegetarian hamburgers, the chain's decision to require its suppliers to employ Grandin's methods was an important step forward.

On an ethical level, Grandin's system encourages slaughterhouse operators to give moral weight to the issue of animal suffering. In this way it shares something important with animal protection theory. All else being equal, it is better for an animal to be killed in a manner recommended by Grandin that it would be for it to die according to a method of slaughter which gave no weight to the animal's suffering, such as killing it with a sledgehammer, an approach still used in parts of the developing world. Although sledgehammers have not been used in American slaughterhouses since the 1950s, slaughter continued to be carried out with little if any regard for the animals' welfare even after this time (Singer 1990, Warrick 2001). Grandin's system has raised awareness regarding food animal welfare and reduced their suffering. Despite its flaws at an ideal level, in the second-best world we actually inhabit, Grandin's method of slaughter has ultimately been a force for good. If it is not as good as no longer eating animals, its superiority over traditional slaughter is still worth recognizing.

CHAPTER 6

What Is It Like to Be a Chicken?

Introduction

A popular view holds that no wrong is done to an animal when it is killed painlessly. As we saw in the last chapter, such a view ostensibly underlies the design for Temple Grandin's system of human slaughter. It finds some support in the work of Peter Singer in cases when the animal in question is merely sentient, although the ramifications of such support are limited by Singer's recognition that more than one animal species have mental capabilities that extend beyond sentience. The notion that at least one common food animal is not harmed by death was also suggested by David Velleman's remark, quoted in Chapter 2, that a cow cannot conceive of itself through time, and so cannot care about its future. Despite the differences between their various views, Grandin, Singer, and Velleman all converge on the thought that painlessly killing at least some animals can involve no moral wrong.

In arguing against this view in the previous chapter I tried to show that it is less philosophically attractive than the time-relative interest account, which does see death as a harm to merely sentient creatures. In arguing for the philosophical superiority of my theory I granted the truth of Singer's views concerning which food animals are likely to have cognitive abilities beyond sentience. As we saw, Singer's discussion left open the possibility that chickens are merely sentient. I framed the discussion this way to make clear that the case against anti-speciesist versions of humane slaughter does not depend on revisionary claims about the cognitive abilities of food animals. On the TRI account, they are harmed even if their mental abilities are as limited as is commonly thought.

But are they so limited? I want to suggest that in the case of at least one widely consumed species, their cognitive abilities are in fact more extensive than is generally recognized. In particular, *Gallus gallus domesticus*, the domestic fowl, has cognitive abilities that amount to a primitive form of self-consciousness. I again focus on chicken because they are the most widely consumed land animal: over 8 billion chickens were killed for consumption in the United States in 2015, compared to approximately 28 million cows and 115 million pigs (USDA 2016, 2017). As humane slaughter systems to date have excluded fish, the debate over the ethics of humane slaughter is primarily a debate over the ethics of killing birds. This debate, however, is occurring at a time when new findings in avian brain science are revising our understanding of their mental abilities. If chickens have mental abilities that place them in an intermediate status between merely sentient beings and persons, this will have bearing not only on the arguments I examined in the last chapter, but on *any* attempt to justify slaughtering them, including attempts that employ Singer's or other versions of protectionist theory.

The revised understanding of chicken cognition thus poses a separate challenge to protectionist arguments for humane slaughter, one that is independent of the challenge posed by the time-relative interest account. Outlining this challenge requires bringing sophisticated moral concepts to bear on cutting-edge neuroscience. On a conceptual level, it requires noting just what primitive self-conscious amounts to, and why its conceptual adoption will mark an improvement for any moral theory that recognizes only two levels of standing. The version of primitive self-consciousness best deployed in this regard, I argue, is one that innovates on previous versions, suitably revised so as to include a primitive ability to conceive of oneself through time.

Primitive Self-Consciousness

Many moral theories divide entities with moral standing into two categories, persons and non-persons. Theories differ in how they distinguish persons from non-persons, but roughly speaking, it is common for personhood to be extended to entities that are not merely conscious but self-conscious. Non-persons with moral standing, by contrast, are typically merely conscious, or sentient. To be sure, there are theories of animal ethics that deny that self-conscious entities warrant a greater degree of moral standing than merely sentient ones.[1] But the counter-intuitive implications of such theories can be seen by considering their implications

[1] Of such theories the most sophisticated is perhaps that of Sapontzis, who accepts the implication regarding lifeboat scenarios described in the immediately following sentences of main text (1987: 221).

for lifeboat scenarios in which, all else being equal, we must choose between saving an ordinary adult human and a dog. On such approaches, the greater cognitive abilities of the human afford no reason to save her life over that of the animal. It is perhaps unsurprising that the most influential theories of animal ethics do in fact ascribe higher moral status to self-conscious beings, which seems a reasonable distinction to draw (Singer 1990: 5–6, Regan 2004: 351).[2]

Yet there is an intermediate category occupied by entities that exhibit something in between full self-consciousness and mere consciousness, namely, primitive self-consciousness. The concept of primitive self-consciousness has been outlined by José Luis Bermúdez (1996). I shall be defending a version of primitive self-consciousness that differs in an important way from Bermúdez's version, but it will be helpful to begin with an outline of Bermúdez's account. I will call Bermúdez's version of primitive self-consciousness the type-one version to distinguish it from my own, type-two, version. Bermúdez's analysis involves not only primitive self-consciousness itself, which is a conceptual category, but a detailed empirical discussion of one type of entity, human infants, whom Bermúdez argues exhibit primitive self-consciousness. These two features are closely intertwined in Bermúdez's discussion, and it will be helpful to take note of his empirical claims in order to illustrate what Bermúdez's pioneering version of primitive self-consciousness involves.

Type-one primitive self-consciousness has been put forward as involving three central features. Shaun Gallagher itemizes them in a summary that builds on Bermúdez's account:

1. A body schema, defined as a subpersonal control mechanism that, in an immediate and close to automatic fashion, provides the capacity to control posture and movement.[3]
2. A differentiation between self and other.

[2] The fact that a deontologist such as Regan accepts that there can be situations in which it is acceptable to sacrifice a conscious entity to save a self-conscious one should remind us that such a view is not necessarily utilitarian. On Regan's view, for example, it is acceptable to sacrifice less cognitively-advanced beings for the sake of self-conscious ones in what he terms "prevention cases," or those in which the only way to prevent some innocent being or beings from being harmed is to harm another (2004: 287). A utilitarian would accept sacrificing one being to save another of higher moral standing in a much wider range of contexts. My account does not offer a final account of how often and under what conditions such trade-offs will be permissible, and entails only that such trade-offs are permissible in prevention cases (lifeboat scenarios).

[3] Body schema is sometimes distinguished from the more complex notion of body image. As Gallagher says of the latter, "body image, when fully developed, is a system of perceptions, beliefs, and attitudes about one's own body. As such the body image involves a reflective intentionality in which one's own body is the intentional object" (1996: 133). Bermúdez does not distinguish between body schema and

3. A recognition that the other is of the same sort as oneself (Gallagher 1996: 133–4).

In order to better understand these three features, it is helpful to note their original use, which was to argue for a revised view of the cognitive abilities of newborns.

A longstanding view in the child development field has held that infants under 8–12 months are not capable of so-called invisible imitation, or imitation of another individual's movements using parts of the infant's body that are invisible to it, such as its face. According to Piaget and other influential theorists, this was because the infant did not have a body schema, understood as a proprioceptive awareness of its own body. Proprioceptive awareness in this usage refers to an awareness of the location of one's hand when it is behind one's back or an awareness that one is smiling without looking in a mirror.

Studies begun in the 1970s challenged the traditional view. They saw infants as young as one day old engage in imitative behaviors involving facial acts, such as mouth opening and tongue protrusion. An experimenter would perform the action while the infant had a dummy in its mouth. After the dummy was removed, the infant would repeat the modelled action. Given that this was impossible on the traditional understanding, the experiments seemed to show that very young infants had several attributes the traditional picture overlooked.[4]

One was a body schema (Gallagher 1996: 133). This was because awareness of the gesture to be imitated and awareness of the imitating action occur in different sensory modalities (Bermúdez 1996: 389). A visual stimulus is followed by an action involving a part of the infant's body, the face, which the infant cannot see, but is rather aware of due to its proprioceptive awareness. The ability to "perceive and act upon intermodal equivalences" in this way requires the existence of some sort of body schema, because such a schema is a basic prerequisite of the integration of perception and motor commands (390).

Bermúdez argues that the experiments also demonstrate that an infant is able to distinguish between itself and another of the same sort as itself. His claim here has to do with the nature of imitation, which he takes to presuppose a rudimentary grasp of the self-other

primitive body image, but Gallagher persuasively argues that the former is all that primitive self-consciousness presupposes (ibid.).

[4] The experiments in question are Meltzoff and Moore (1977); Meltzoff and Moore (1983); Field et al. (1982).

distinction. The infant, Bermúdez suggests, has to have at least a dim comprehension of the fact that the entity it observes is one whose action it can copy:

> Not only do these infants have to grasp the intermodal equivalence of, say, a perceived tongue protrusion and their own act of protruding their tongue, but they also have to grasp that the perceived tongue protrusions *is the sort of thing that they can imitate*, as opposed, for example, to the movement of a nipple, to which they respond by sucking rather than by trying to imitate it (1996: 390).

When the infant sees a nipple, it responds by attaching itself to it. When it sees an experimenter sticking out her tongue, by contrast, it responds by sticking out its own tongue. This second type of reaction would appear to presuppose an awareness that the experimenter is doing something of which the infant itself is capable. Or as Bermúdez puts it, the infant's level of awareness "seems to involve a grasp that the experimenter is a being like themselves, in at least the sense that they are constructed in more or less the same way" (Bermúdez 1996: 390). If that is the case then the infant is aware, however dimly, that the experimenter and it are different entities with some sort of affinity. That would appear to demonstrate some minimal awareness of the self-other distinction, and an awareness that the other is somehow similar to itself.

If these conditions hold, Bermúdez argues, then the infant has a crude form of self-consciousness. A body schema for example would appear to be a kind of self-awareness, if only the awareness that one is an embodied or physical being. This is a primitive form of self-consciousness that the more sophisticated or full versions draw on, as a body schema informs our sense of what our possibilities for action are (ibid.). Similarly, the infant's imitative response would suggests that it is aware of a similarity between two entities, one of which is itself. This awareness again may be quite rudimentary, but its content concerns the nature of the infant itself. In particular, it is aware of possessing at least one possibility for action compared to the entity it observes. In taking that entity's actions as inspiration for its own, it thus appears to manifest some limited awareness of self.

Before completing our exegesis of type-one primitive self-consciousness, it will be helpful to note the relationship between primitive self-consciousness and its more full-blown counterpart. Exactly how is the first form of self-consciousness a more primitive version of the latter?

To answer this question we need to note two features of self-consciousness proper.

The first is a capacity for reflexive self-reference, as when we refer to ourselves using the first-person pronoun. This capacity to single ourselves out in a reflexive way makes possible many of the high-level cognitive actions we associate with full self-consciousness. According to Bermúdez, they include the "capacity to make and reflect upon plans for the future, to form second order desires, and to fit together autobiographical memories into the narrative of a single life" (1996: 391) Reflexive self-referencing has other components, but one important aspect is that it is a tool self-conscious subjects can use to distinguish themselves from other self-conscious subjects. Using the first person pronoun is a way of "delimiting oneself as the object being spoken or thought about," as distinct from other self-conscious beings one might think or speak about (ibid.).

The second noteworthy feature of full self-consciousness is essential self-knowledge, or knowledge of the sort of being that one is. In the case of adults, this might include the knowledge that one is an embodied human being, with powers of rationality, who is a member of a community with other such beings, etc.

Bermúdez's central claim is that infant imitation behavior displays a primitive form of both of these features of full self-consciousness. The understanding of the distinction between self and other that an infant displays in imitating an experimenter's behavior is an early form of the much more complex distinction between self and other that occurs in full self-consciousness through reflexive self-reference. Similarly, the infant on Bermúdez's account displays a rudimentary form of essential self-knowledge by grasping that it is a being like the experimenter. Another way to say this is that the infants manifest a certain degree of essential knowledge and employ it to govern their interactions with the experimenter. This was suggested by their awareness that the acts they see are ones they can perform. Such an awareness is a kind of self-knowledge. To be sure, it is a far less developed form of self-knowledge than the full-blown version, which is entirely conceptual and propositional, while the baby's version is more akin to knowing-how than knowing-that. Nevertheless, the two forms of self-knowledge are on the same developmental continuum. And so the baby's self-knowledge stands in the same relation to an adult's as its crude grasp of the self-other distinction stands in relation to a more sophisticated version that is manifest through reflexive self-reference.

To sum up our exegesis of Bermúdez, what is important on his analysis is not an infant's capacity for imitation qua imitation. Imitation rather bespeaks some minimal level of self-awareness.[5] If we conceive of self-consciousness as a spectrum, with full self-consciousness at one end, and a much more minimal version at the other end, the infant seems conscious of itself enough to be just inside the boundary of the primitive level of self-consciousness. That is not very impressive when we have full self-consciousness in mind. But it is more than would be true of an entity that was conscious, with absolutely zero degree of self-consciousness, not enough even to know it could imitate what it was seeing.

Now one possible objection to this interpretation of infant imitative behavior is to say that what is happening is a reflex. On such a view, the infants are not exhibiting any degree of self-awareness, but are rather exhibiting an "innate releasing mechanism," or a pre-programmed response to perceived events exhibiting a common structure (Bermúdez 1996: 392). This view, however, is normally taken to apply to actions that occur without the involvement of memory: rather the stimulus arrives and is met with the pre-programmed action in question. In the case of the infant experiments, however, the infant has a dummy in its mouth when it sees the model behavior and only later imitates it. This means there must be a stored representation of the model's facial gesture that triggers the infant's imitation, in a manner that seems distinct from how innate releasing mechanisms are usually understood. Bermúdez also notes that the infants in the experiments corrected their responses over time, and trial and error activity is normally thought not to be characteristic of reflexive behavior. If so this provides an additional reason to reject the reflex interpretation.

Derived Moral Significance

This concludes our outline of type-one primitive self-consciousness. As we will see, the version outlined next has an important difference with Bermúdez's account. But an important feature both accounts share is a common source of appeal. Both make appeal to the principle of derived moral significance, or the DMS principle for short. It takes the following form:

> If a particular feature or property is deemed to confer moral significance upon a life that has it, then any primitive form of that feature or property

[5] Bermúdez also suggests that an infant's initial act of imitation activates its primitive self-consciousness. See the following.

> will also confer moral significance, although not to the same degree (Bermúdez 1996: 383).

Such a principle entails that if a property or attribute is morally salient, then any entity which has a primitive form of that attribute also possesses a corresponding degree of moral significance. Applied to the property of self-consciousness, the DMS principle will entail that any entity that is primitively self-conscious is also possessed of a morally salient trait, albeit not to the same degree as an entity that exhibits full self-consciousness. The primitively self-conscious entity, importantly, possesses this morally significant trait independently of any other morally significant attributes it might possess (such as, for example, a capacity for suffering).

In order to better understand the DMS principle, it will be helpful to note two claims it does not entail. It does not entail that an individual being who is currently conscious, and in the normal course of events will become self-conscious, possesses some additional degree of moral standing in virtue of its potential for self-consciousness. An infant for example is conscious and in the normal course of events will become a person, and there have been moral theories which have sought to ascribe moral significance to this potential personhood. The DMS principle is not like this: it makes no reference to potentiality. Rather it ascribes value to primitive self-consciousness insofar as it is an *existing* feature of a living thing, and only insofar as it is an existing feature. The claim is that an account which makes the existence of a particular trait morally significant "ipso facto makes the existence of primitive forms of that feature morally significant" (Bermúdez 1996: 385). Potentiality of any kind never enters into the analysis.

A second claim the DMS principle does not entail is that if an entity exhibits a primitive form of a morally significant property, then it will necessarily go on to possess that property in a fuller form. It is rather possible for the proportionate moral significance which the principle generates to apply to an entity that will never possess the full version of the attribute in question. We can imagine a human being who is cognitively disabled, such that his course of development does not follow the normal path leading to full self-consciousness. Instead he reaches the stage of primitive self-consciousness and stops there. The DMS principle would ascribe moral significance to this limited amount of self-consciousness, albeit, of course, less significance than that possessed by a fully self-conscious person. In a similar way, the DMS principle can also apply to a member of a species which does not have the capacity for full self-consciousness, but does exhibit a primitive form of self-consciousness.

Why adopt the DMS principle? One reason is because it avoids classifying sentient beings into one of only two categories of moral standing. As mentioned previously, moral theories frequently divide entities with moral standing into persons and non-persons. Let us refer to such two-level theories with the shorthand term "personhood theories." It is common for personhood theories to maintain that every sentient being below the level of a person has the same degree of intrinsic moral standing (I say intrinsic in order to set aside moral issues having to do with personal relationships and social roles).

Personhood theories do not seem able to capture the notion that there could be two sentient beings, neither of which was a person, but which nonetheless occupy different positions of moral worth. A bird and a fish, for example, may both fall short of personhood. But this fact, by itself, does not necessarily entail that they both must possess an equal degree of intrinsic moral standing. It is at least possible that the bird, which may have cognitive abilities beyond that of the fish, deserves a heightened position of intrinsic moral standing. Such a position may fall short of that of a person. Nevertheless it could still make a moral difference, such that if we had to choose between killing the bird or the fish, it would be less wrong to kill the fish. Given the diversity of living things, it seems an advantage of a theory of moral standing if it can at least admit such a possibility in principle, without presuming in advance that any particular animal does in fact occupy a position of worth below personhood but above that of a merely sentient creature.

A theory that admits at least the possibility of an intermediate status between sentience and personhood will respect the DMS principle and be more nuanced and subtle than a theory that lacks any intermediate stage. Given the diversity of animal life, and the fact that all theories of moral standing ascribe moral significance to the intrinsic properties of living things, it is reasonable to see a theory that is better able to reflect complexity in such properties as a better theory, all else being equal.

This last point might be thought to prove too much. If a three-level theory of moral standing is better than a two-level theory, we might wonder, does it not stand to reason that a four-level theory is better still, a five-level theory even better than that, and so on? If we are to increase the levels of moral standing from two to three, on what consistent grounds can we stop short of increasing it beyond three?

The limited cause for concern here bears noting. The concept of primitive self-consciousness does not add a new property to the list of those that are relevant to determining inherent moral standing. There remain only two, sentience and self-consciousness. The deep structure of our thinking about moral standing is preserved rather than overhauled. So there are no grounds to fear that primitive self-consciousness will somehow give rise to a new level of moral standing that applies to entities that are not even sentient. Similarly, there are no grounds to fear that it will give rise to a new level of moral standing higher than that of persons, a level that would be illustrated by aliens, androids, or other creatures of science fiction with mental capabilities that have the same relation of sophistication and complexity to self-consciousness as self-consciousness has to mere sentience. Because primitive self-consciousness applies only to entities with attributes that fall in between sentience and self-consciousness proper, its ability to generate new levels of moral standing is confined only to intermediate cases.

We might still worry that ascribing moral significance to primitive self-consciousness will oblige us to recognize more than one intermediate stage of standing between sentience and full self-consciousness. But what exactly is the worry here? It seems a stretch to say that ascribing import to primitive self-consciousness *entails* also ascribing import to some further intermediate stage in between it and self-consciousness proper. The presence of any degree of self-consciousness is metaphorically the difference between zero and one, a difference one can ascribe significance to, while stopping short of the claim that every additional degree of self-consciousness is equally significant. It is the presence of something not found at all in merely sentient creatures that makes primitive self-consciousness a crucial threshold, in a way that does not necessarily carry over to creatures that possess a degree of self-consciousness in between the primitive and full levels. While one might argue for such a highly graduated or four-level theory of moral standing, it is not strictly entailed by the principle of derived moral significance. Even if the present account did entail a four level theory, this does not seem a fatal problem. Such a theory would be complex but manageable. The real concern would seem to be a theory of many degrees of moral standing. Intuitively, though, it seems more plausible to posit a single threshold of significance between primitive and full self-consciousness, rather than to endlessly divide self-consciousness into every more precise gradations of significance. Given this, and absent any argument for four or more levels of standing, it seems reasonable to proceed making the case for three.

Type-Two Primitive Self-consciousness

I turn now to outlining a notion of primitive self-consciousness that incorporates an important new element. Whereas Bermúdez focuses on the possession of a body schema, an ability to distinguish between self and other and an awareness that another is similar to oneself, I employ a notion of primitive self-consciousness that additionally includes some minimal ability to conceive of oneself through time.

Why adapt primitive self-consciousness this way? The ability to conceive of oneself through time is often cited as a morally relevant aspect of self-consciousness (Tooley 1983; Singer 1993; Velleman 2000). The precise role this ability plays differs from one theory to another, but a common feature is its connection to the ability to form desires about the future. A person can form forward-looking desires, such that if the person is killed, it will prevent the attainment of those desires. Hence a person is deprived of something by death in a way that is not the case for an entity that is unable to form desires or expectations regarding its future.

It is possible to conceptually separate the ability to conceive of oneself through time from the aspects of self-consciousness discussed by Bermúdez. One might imagine for example a hypothetical creature that conceives of itself through time but is unable to distinguish between itself and another, as it has never encountered another sentient being. In regard to real entities, however, it seems unlikely that an animal will possess an ability to conceive of itself through time but completely lack the aspects of primitive self-consciousness highlighted by Bermúdez. Rather, it seems in nature that cognitive ability is a "package deal," in that an entity is more likely to have some primitive ability to form expectations regarding the future if it possesses the features of primitive self-consciousness highlighted by Bermúdez.

If an ability to conceive of oneself through time and an ability to form desires regarding the future are morally salient aspects of full self-consciousness, then the principle of derived moral significance entails that a primitive ability to exercise such features will warrant an equivalent degree of moral significance. Applied to the ethics of killing, the ability of persons to form desires regarding the future is typically taken to warrant recognizing their possession of a right to life, or something like it (Tooley 1983; Singer 1993; Velleman 2000). The DMS principle suggests that if an entity is primitively self-conscious, it will not possess this same right. On the other hand, such an entity will have some interest in continued existence. It will be the kind of thing that can be harmed by death, in that it is deprived of something by being killed. I will try to outline what moral entitlement best corresponds with this kind of primitive self-consciousness, but the point at hand is that a primitive ability to conceive of oneself though time is a morally salient attribute.

With these considerations in mind, let us now ask: do chickens possess primitive self-consciousness? Our concern here is with primitive self-consciousness not as Bermúdez defined it, but the type-two version that includes a primitive ability to conceive of oneself through time and form expectations regarding the future. An animal is more likely to possess this ability if the animal possesses the aspects contained in type-one primitive self-consciousness. Is it conceivable that chickens might not only meet Bermúdez's criteria of primitive self-consciousness, but also have some sense of time and the future?

It is relatively uncontroversial to ascribe greater cognitive abilities to chickens than to newborns.[6] For example, the ability to retain recognition of partly hidden shapes, an ability possessed by two-day-old chicks, does not emerge in humans until 4–7 months (Lea et al. 1996). Tests of two-day-old chicks' abilities to recover fully occluded objects have also found that they mastered some aspects of stage four of the so-called Piaget Scale of object permanence, a level human infants do not begin to reach before three and half months of age (Regolin et. al. 1994, Regolin et. al. 1995; Baillargeon and DeVos 1991).[7] Such findings are perhaps unsurprising when one considers the environment chickens are born into. As avian cognition researcher Giorgio Vallortigara points out, "recognition of a partly occluded mother would be useful when an organism can move by itself to rejoin the mother and thus reinstate social contact; that is the case for the highly precocial young chick, but not for the highly altricial species such as the human newborn" (2004: 62). The findings just mentioned do not seem to presuppose the existence of any form of primitive self-consciousness. They should nonetheless open us to the possibility that,

[6] See Marino (2017) and Garnham and Løvlie (2018) for overviews of current findings regarding the cognitive abilities of chickens.

[7] See Vallortigara (2004: 61–2, 73–4, 80) for summaries of both experiments and the relative performance of chicks and infants. Vallortigara also mentions a third study testing animals' ability to distinguish between different corners of a rectangular cage using only geometric information about the cage's layout. The animal was placed in the cage when food was located in a corner and made to disappear. After the animal was disoriented, it was allowed to search for the previously present food. More than one animal species was able to distinguish between different corners based only on the rectangular shape of the cage. According to Vallortigara, "the performance of birds in these tasks is identical to that of rhesus monkeys (and human adults) and clearly surpasses that of rats of human infants" (2004: 80).

here as elsewhere, chickens may be more cognitively advanced than newborns.

Is there any direct evidence for type-two primitive self-consciousness in chickens? Consider a study that investigated the behavior of mother hens (Nicol and Pope 1996). The hens were exposed to two different food sources that had been treated with food coloring. One colored food was edible while the other was not. The particular colors used varied from hen to hen, but for simplicity's sake we can imagine a hen choosing between a plate of yellow food pellets, all of which were edible, and a plate of blue ones, all of which were inedible. The hens quickly learned to avoid the unpalatable food. They were later exposed to their chicks participating in a series of feeding sessions in a neighboring pen. Each hen had an even number of chicks, some of which ate food sources with similar color coding: a mother who had been exposed to yellow edible crumbs and inedible blue ones would watch some of her chicks choose between sources with the same color coding. Each mother however was also exposed to a second set of offspring, whose food options had a reverse color coding, with the yellow now unpalatable and the blue palatable. So far as the hens could tell, the second chicks were making a mistake and opting for unpalatable food. The researchers wanted to know if this information would have any influence on the hens' behavior.

They found that it did. Hens are known to engage in maternal food displays, or actions which facilitate the acquisition of feeding skills on the part of their chicks. Such displays include food calls and pecking movements directed toward both food items and the ground. After witnessing the second set of chicks eat the "wrong" food, the hens significantly increased their maternal food displays, engaging in more intense scratching and ground pecking than after witnessing the first group. The researchers concluded that hens modified their own behavior in response to perceived feeding errors by their chicks, seeking to instruct them in what type of food to avoid.

The behavior of the hens and chicks provides as much evidence for type-one primitive self-consciousness in chickens as the child-development studies cited by Bermúdez provide for a similar judgement regarding infants. Recall that Bermúdez noted that a primitive body image was suggested by an ability to integrate visual perception and motor commands. The chicks and hens both do this when they opt for food of one color and not another. (Additionally, the hens accelerate their food displays in response to the visual stimulus of watching their chicks eat.) Insofar as primitive body image is present in an entity that can use its body in a manner sensitive to what it sees, chickens pass the test.

The study also suggests that chickens meet Bermúdez's second and third condition, that of being able to distinguish between self and other, and to recognize that one is an entity of similar kind to another individual. This was the conclusion Bermúdez drew from the infants' ability to engage in imitative behavior. Maternal food displays presuppose a similar ability on the part of both chicks and hens. The chicken researchers note that "newly hatched chicks peck at food and non-food items alike and have to learn which items are palatable and profitable" (1996: 768).[8] Maternal food displays facilitate this type of learning, as the hen pecks at a palatable food item while vocalizing, or picks it up and drops it. In order for the chick to benefit from this type of behavior it has to recognize that the hen is engaging in a behavior, that of pecking at food, that it also can engage in. It has to be at least dimly aware that both it and the hen can eat the same thing.

Similarly, in order for the hen to be concerned about the chicks' perceived food error, she has to recognize an affinity between a chick's action and what she was doing when she ate the unpalatable food. She has to have some minimal awareness that the chick is doing what she did, which presupposes an awareness that the chick is the kind of thing that she is.

Bermúdez argued that the involvement of memory and trial and error each cut against the judgment that the infants are acting on instinct. Both attributes are present in the chickens' behavior. Both the chicks and the hens are able to remember that food of a certain color is unpalatable. After seven days of training, all subjects consistently avoided the unpalatable feed, a behavior they maintained throughout the remaining four days of the experiment, during which less than one out of a thousand pecks were directed at unpalatable food. When the hens watched the chicks eat the "wrong" food and accelerated their display behavior, they did so for one-hour periods. During the first half hour, the hens had no access to food of

[8] That chicks' choices between different colored foods are influenced by observing the distaste responses of conspecifics is further supported by Skelhorn (2011). Sherwin et al. also found that adult hens "developed preferences for novel food by watching the feeding behavior of conspecifics" (2002: 940). Unlike chicks, however, hens did not learn to avoid unpalatable food through observing disgust reactions. The researchers suggest this may be because adults have previous experience with food, and so they make decisions based on this experience rather than their observations of other hens. Nicol and Pope (1992, 1994) found that adult hens who watch a trained hen perform an unfamiliar task become better at performing the task in question. Chickens appear to learn such tasks most readily from high status conspecifics, to whom they pay more attention (Nicol and Pope 1999).

any kind, and had last eaten palatable food ninety minutes previously, and had not eaten unpalatable food for at least 72 hours. When the hens intensified their displays in response to the chicks' perceived mistake, therefore, they were acting on a stored representation rather than an innate releasing mechanism or instinct.

The hens displayed no tendency to intensify their maternal displays in response to watching their chicks eat food of any particular color. They rather changed their behavior whenever they saw the chicks eat the food that was the same color as their own unpalatable food, a color which varied from hen to hen. It was not the color per se that triggered their behavior, in other words, but the memory that food of such color was in their experience unpalatable.[9]

As for trial and error, it too is presupposed by the chickens' behavior, as this is the context in which maternal food displays function. Baby chicks do not arrive in the world knowing what to eat. They learn where to direct their efforts by observing their mother's behavior. As such, they engage in a form of rudimentary trial and error behavior, up to the same level of an infant refining a tongue protrusion or similar facial gesture. In that sense there is just as much evidence cutting against a purely instinctive or reflexive explanation in the case of the animal as in the case of the infant.

Chicks and chickens, just as much as infants, meet Bermúdez's criteria of type-one primitive self-consciousness. Turning now to type-two primitive self-consciousness, is there any evidence to suggest that chickens have a primitive sense of the future? Here we turn to a recent experiment that found chickens can display self-control when choosing between two time-released food sources (Abeyesinghe et al. 2005). The chickens were trained to peck on one of two keys that would temporarily release a quantity of food into a trough after a short delay. Pecking the first key would release the food after a two second delay, and it would stay in the trough for three seconds before disappearing through a drop-away floor. Pecking the second key, by contrast, would release the food after a six-second delay; but it would stay in the trough for 22 seconds. The researchers

[9] The capacity of chickens to make decisions based on stored representations is well-supported by additional studies of chicken memory. Such studies have found that hens and chicks retain a memory not merely of the location but also the content of food caches (Forkman 1998, Cozzutti and Vallortigara 2001); that five-day-old chicks can determine the location of a moving object three minutes after it disappears, a result similar to that found in studies involving primates (Vallortigara et al. 1998); and that one-day-old chicks who observe another chick have an adverse reaction after pecking a bead coated with a bitter chemical learn through observation that they should not peck the bead themselves, and avoid doing so for up to 24 hours after the observation (Johnston et al. 1998; see also Salva et al. 2009) .

labeled the second choice the jackpot choice, as it allowed the chicken a much greater food access.

The experiment was designed to test whether chickens had a capacity for self-control, with self-control understood as the ability to "resist immediate gratification for a later benefit" (Abeyesinghe et al. 2005: 2). The study found that the chickens' tendency to exercise such self-control was "significant and pronounced" (10). Of the seven birds trained to choose between the two options, all seven opted for the jackpot choice over 75 percent of the time. Five of the seven did so over 95 percent of the time.[10] This overwhelming preference for the jackpot option suggests that the birds had some awareness of the short-term consequence of their decision regarding which key to push. As the researchers put it, "Our findings have demonstrated, first, that hens can discriminate between short time periods; second, that there is no absolute cognitive deficit preventing the hen from comprehending later consequences of food choices" (8). The experimenters do not claim to have shown precisely how far into the future hens can perceive: they note that little research has been conducted into this question to date. But the overwhelming preference of the birds for the jackpot food access over the impulsive option or random results does demonstrate an ability to see at least six seconds into the future. As the researchers put it, "although how far into the future hens can mentally travel has yet to be determined, we now know that they are not completely 'stuck in time'" (8, 10).

Consider now a third and final study, which also investigated whether chickens have some perception of time (Taylor et al. 2002). It saw five chickens successfully trained to peck a lit touch screen to obtain a food reward. After the chickens learned how to use the screen, the researchers introduced a time delay into the experiment. After the screen came on, six minutes would elapse before any food could be obtained. The first peck of the screen after six minutes had elapsed would result in the food reward. The chickens were subjected to 150 training trials followed by 100 testing trials. Previous studies on other species have found that some animals

[10] The researchers noted that different results obtained when the amount of time required for the jackpot choice changed. When the jackpot choice was changed to involve a 22-second delay to a 22-second food access, for example, six out of eight chickens opted for the two-second wait for a three-second feed access over 95 percent of the time. The researchers suggest that the different results may be due to the birds discounting the value of a more temporally distant reward (ibid., 6). By contrast other studies have found that chickens can anticipate rewards 15 seconds (Zimmerman et al. 2011) or 22 seconds in the future (Moe et al. 2009). For a critical review of time preference and time discounting studies of animals see Hayden, who suggests that "animals may be much more patient than is widely believed" (2016: 39).

respond less at the beginning of the delay period and more frequently after it has elapsed. "The advantage of this behavior is that the food reward is obtained reliably and the effort used in obtaining it is reduced, due to the low rate of response at the beginning of the interval. The frequent response near the end of the interval allows food to be obtained as soon as it becomes available" (42). The results of the chicken experiments found that they too pecked the screen comparatively less during the first two minutes of the time delay, and accelerated their pecking so that it peaked in frequency between six and 7.5 minutes. The researchers' conclusions are tentative, among other reasons, because they were only able to perform a limited number of trials over a two-week period. Nevertheless, they take the results to raise the possibility that hens do have a limited perception of time. "The results indicate that domestic hens may have the ability to estimate the time to reward when given a reliable signal such as that given by the touch screen several minutes in advance" (49).

This experiment and that previously involving self-control are similar to that involving maternal feed displays, in that the behaviors under investigation were learned through trial and error. In both time experiments, the chickens had to be trained to use an experimental device. The behaviors in both time experiments equally presupposed some stored representation of what the animals had learned: they had to retain some memory of how the equipment worked from their training, and apply that knowledge to their use of it in the experiments proper. These considerations cut against the notion that the birds' activity in either experiment was purely instinctive. In the first experiment, the instinctive action would be to peck at either key with random frequency, with no sign of deliberation between the two. In the second, an instinctive response would be less likely to vary across time. Rather the bird would again peck in an indiscriminate way. The time experiments thus seem to demonstrate behaviors that can no more be characterized in reflex or instinctive terms than the behavior of the infants in the experiment Bermúdez discusses or that of the chickens in the maternal food display study.

The time-perception studies suggest that chickens have a primitive aspect of the dimension of self-consciousness that involves a perception of oneself through time. Particularly in regard to the first time-study, the chickens were able to choose between two options that involved different rewards at different points in the future. As ever with primitive self-consciousness, the ability at hand is not very impressive when compared to its full-blown equivalent. But the perception of time in relation to their future is present nonetheless. This suggests that a reason that has sometimes been given for the permissibility of killing chickens, their lack of any awareness regarding the future whatsoever, may be based on a factually incorrect claim.

If chickens have a primitive awareness of the future, we might wonder how it was the case that this has traditionally been denied. For a century, scientists thought that the neocortex was a prerequisite for higher cognition. As the neocortex is found only in mammals, research into the cognitive abilities of other animals was considered pointless, as birds were assumed to lack the brain feature necessary for malleable, that is, non-instinctive, behavior. However, in recent decades it has emerged that functions performed by the neocortex in mammals are performed by different parts of the avian brain. This has caused the Avian Brain Nomenclature Consortium, an international group of neuroscientists, to overhaul the terminology that scientists used for 100 years to identify different parts of the avian brain (Avian Brain Nomenclature Consortium 2005).

According to the consortium, the new picture of the avian brain "sets the stage for a re-evaluation of the cognitive abilities of birds" (Avian Brain Nomenclature Consortium 2005: 156). Precisely such a re-evaluation is represented by the work of the animal scientists I have cited. The new picture of avian cognition they suggest calls into question the popular notion that primates occupy a position of exceptional cognitive standing among animals. For example, Nathan Emery and Nicola Clayton, an animal behaviorist and experimental psychologist, respectively, at Cambridge University, ask: "Why have primates achieved this special status when the evidence suggests that species of birds have cognitive abilities that are equal to or more sophisticated than have been demonstrated for primates?" (Emery and Clayton 2004: 36–7) A similar viewpoint is taken by Lesley Rogers and Gisela Kaplan of the Centre for Neuroscience and Animal Behavior at the University of New England. They note that recent animal cognition research "challenge[s] the concept that primates are special and even the view that the cognitive ability of apes is more advanced than that of non-primate mammals and birds" (2004: vii). The same point is made in regard to chickens by Lucia Regolin, a psychologist at the University of Padua, and three co-authors. Regolin and her colleagues point out that domestic chicks have "striking abilities to maintain working memories of the location of biologically attractive objects that parallel or outperform those of most mammals, including primates" (Regolin et al. 2005: 855). Taken together, the new image of the avian brain ask us to take seriously the possibility that birds such as chickens possess mental abilities no less sophisticated than mammals. That scientists have only recently explored the possibility that avian brains can perform functions traditionally associated with the neocortex would also explain why the revised view of avian cognition has only recently emerged.

The Benefit-of-the-Doubt Standard

Moral conclusions that presuppose revisionary scientific claims would seem necessarily tentative. We are dealing with empirical findings that could potentially change. Nevertheless, I believe the aforementioned experiments show this much: it is reasonable to ascribe chickens whatever moral standing accompanies type-two primitive self-consciousness in a manner that gives them the benefit of the doubt.

What justifies giving the birds such a benefit? The reason has to do with a danger that comes with determining which entities possess any morally salient attribute. The danger is one of causing harm by failing to recognize the attribute's presence in a given entity. Investigating whether chickens possess primitive self-consciousness is not like investigating what colors their feathers are. When it comes to a morally irrelevant feature such as color, we are not at risk of causing harm by employing an overly strict method of classification. But when it comes to a morally salient feature such as primitive self-consciousness, if the birds possess it and we fail to recognize this through an overly scrupulous standard of proof, we could potentially do something morally wrong. For example, we might kill them in a wider range of contexts than their cognitive abilities and corresponding moral status permit when properly understood. The best way to avoid this danger is to err on the side of recognizing primitive self-consciousness.

The legitimacy of a benefit-of-the-doubt standard would seem influenced by whether or not the evidence for the attribute under investigation is above some minimal threshold, a threshold that makes it at least possible that the attribute is present. For example, suppose an environmental ethicist were to argue that we should treat stones as persons. Imagine the environmentalist urged this view on us because it was at least open to doubt, and we should give stones the benefit of the doubt. Surely this would be a weak argument, given that there is no evidence that stones exhibit personhood.

When it comes to type-two primitive self-consciousness, by contrast, there is enough evidence to make its presence in chickens an open question. That should cause us to make claims about the individual's moral standing in a manner that errs on the side of higher, rather than lower, moral ranking. This is not because we want to commit ourselves to unsupported views regarding animal cognition. Rather, it is based on a recognition of the difficulty of knowing what goes on inside the mind of creatures we cannot speak to directly. In the case of infants, for example, it seems reasonable to err on the side of extending them the protection of morality based on what we do know about them. The same interpretive stance should apply in our dealings with animals which, if anything, appear to have cognitive abilities beyond those of newborns.

The Moral Significance of Type-Two Primitive Self-Consciousness

If chickens do possess type-two primitive self-consciousness, what moral significance follows from this? One's answer to this question will inevitably be influenced by one's ethical theory. As we have previously noted, most moral theories one encounters today give some weight to the equal consideration of interests. Equal consideration need not apply at the level of judging individual actions: one could also employ it to determine what type of rule to adopt. Similarly, as we have seen, a commitment to equal consideration can be found in deontological and utilitarian theories alike. While there is ongoing debate as to what equal consideration precisely entails, the basic notion of equal consideration is widely endorsed. It is therefore worth asking what difference type-two primitive self-consciousness makes when combined with some version of the equal consideration of interests principle.

One difference is that it no longer is plausible to maintain that chickens are not harmed by death, even on approaches such as Singer's.[11] As we saw in the previous chapter, Singer's account tentatively permitted the painless killing of nonpersons when they were replaced by other nonpersons. The case for killing them painlessly presupposed that they have no welfare interest in continued existence, and so are not deprived of anything when they are killed, so long as it is done painlessly. However, this assumption does not hold for entities that possess type-two primitive self-consciousness, who do have at least some interest in continued existence, yet are not persons.

[11] The widespread view that chickens are not harmed by death is my target in this paragraph. But note that my discussion of primitive self-consciousness in chickens will also affect their standing in ethical theories, such as my TRI account, which hold that merely sentient beings can be harmed by death. The presence of primitive self-consciousness will increase the *scope* of the harm caused by death, and so make killing it that much worse than killing a merely sentient being.

Although that interest is not on the scale of that possessed by persons, it is nonetheless significant, as it means that a primitively self-conscious entities has interests that need to be taken into account when we are deciding whether or not to kill it. Importantly, we need to do so not merely with a time-relative interest framework, but when employing any theory that endorses equal consideration. In order for killing them to be justified, some morally significant greater interest will need to be advanced by their death. This means that while chickens can be killed for a wider variety of reasons than persons, there are some instances in which it is wrong to kill them, even when they die painlessly.

How might we characterize such a level of moral standing? I believe it is best formulated in terms of what can be termed a "slow-to-kill" entitlement. Such a degree of standing is one that would permit killing a primitively self-conscious entity in a wider range of contexts than a person. If my car were out of control on a wet road for example, and the only way to save a person's life was to swerve into a different lane where a chicken happened to be, the slow-to-kill stance would license killing the bird.[12] A person has greater cognitive abilities, including a greater ability to form expectations regarding the future. When comparing a life for a life, the person will lose more extensive welfare-interests when it is killed than will a primitively self-conscious entity, whether human or animal. Thus there would also be grounds to sacrifice the life of a primitively self-conscious human orphan in similar cases where I am forced to choose between killing such a being and killing a person. The person's interest in continued existence will outweigh the interest of the merely primitively self-conscious being, animal or human, has in its future. We can therefore legitimately say that saving the life of the person advances a greater moral interest, with relative ranking determined according to the equal consideration standard.

By the same standard, however, the most common reason for killing chickens – so we can eat them – does not measure up. The animal loses its life when it does not really need to. For we could easily live without chicken in our diets. For this reason, there is not a real need at stake. Rather a moral wrong occurs when we kill a primitively self-conscious entity for such a purpose. The advertising that directs us to eat chickens because they are finger lickin' good, for example, would seem to reflect a morally trivial reason to kill them. That we enjoy how an animal's body tastes is not

[12] This case is similar to Lever cases, discussed in Chapter 4, insofar as the death of the chicken is not the means by which I save the life of the person.

sufficient grounds to override the animal's interest in continued existence.[13]

This conclusion can be supported by way of a thought experiment mentioned in Chapter 2. It concerned the members of the affluent and otherwise egalitarian society who demand that a family on their block be denied the opportunity to build a home, so that that family's plot of land be kept open for collective use as a park. As we have seen, there are moral theories that would endorse taking the last person's lawn, utilitarianism being the leading example, in either its preference-based or hedonist formulations. This gave rise to the criticism that utilitarianism has an inadequate conception of moral equality, insofar as it grants weight to preferences or desires that themselves fail to respect equality. Hence Kymlicka's criticism that the homeowners fail to take into account what rightfully belongs to the last family, a failure to recognize their legitimate moral interests. "Hence prejudiced and selfish preferences are excluded from the start, for they already reflect a failure to show equal consideration" (Kymlicka 1990a: 42).

Kymlicka's understanding of equal consideration is intuitively plausible. But as we have seen, if we accept Kymlicka's restriction against selfish and biased preferences, then the desire to eat an animal that is harmed by death will not be granted any moral weight.[14] This conclusion is considerably

[13] One writer, Jean Kazez, has argued that the pleasure of taste can constitute a necessary ground to harm animals (2017). Kazez's ethical framework, however, "doesn't rule out imposing greater bads on animals to avoid smaller bads for ourselves" (10). Kazez provides no argument for this framework. It is also incompatible with the equal consideration of interests, which as I have argued is a principle any minimally plausible ethical theory must embrace. Kazez also suggests that some people experience a significant loss of food enjoyment when they give up meat, citing the fact that some people revert to omnivorism after they try going vegetarian or vegan. However, research on people who backslide on a dietary code suggests that such failure can be mediated by the individual's ethical beliefs regarding animals and food (Rothgerber 2015; Hodson and Earle 2018). The mere existence of backsliding therefore does not show that it is caused by a serious loss of food pleasure. It could be due to the fact that the individual is not fully committed to the dietary code in question or lacks social support. Finally, the fact that meat eaters cannot always tell the difference between meat and plant-based imitation meat products calls into question the degree to which giving up meat would necessarily involve a loss of pleasure from taste (Lynch 2008; Stump 2014).

[14] While I am here operating in an explicitly anti-utilitarian framework, I believe the avian cognition research I have cited will also have similar ramifications for some versions of utilitarianism. Singer in his utilitarian mode has argued that we have a substantial obligation to avoid killing entities that are capable of "conceiving themselves as distinct beings with a past and a future" (1993: 131). If I read Singer correctly, to the degree that an entity possesses this ability at all, even to a degree that falls short of that of a normal adult human, then "the case against killing [it] is strong" (1993: 132). By that standard, when the new avian cognition research is analyzed in Singer's terms, the case against killing chickens is strong. For a version of utilitarianism that posits a third level of moral standing between merely sentient beings and persons, see Varner (2012). Varner's intermediate level is that of a "near person."

strengthened in the case of animals that possess primitive self-consciousness, such as chickens. This is because the preference to eat such an animal is one that does not take account of the full strength of the bird's interest in continued existence.

This strong interest is entailed by the bird's ability to form expectations regarding the future and a conception of itself through time, both of which are destroyed when it is killed. Given that the bird suffers a greater deprivation when it is killed than a person does when she must eat an alternative food source, especially an alternative source that is easily available, a preference that takes equal consideration into account at the level of its formation will be one that does not destroy the animal's strong interest for a purpose as trivial as eating it for reasons other than sheer necessity. There are thus even stronger grounds than those presented in Chapter 2 to say that the desire to eat a chicken exhibits a selfish disregard for the interest of the consumed animal, a refusal to recognize what is rightfully its. As before, there is no interest to weigh against the animal's interest in continued existence, as the interest in satisfying the desire to eat such an animal is never recognized as a morally legitimate interest to begin with.[15] But the scope of the harm done to the animal is now revealed to be more significant than previously thought.

Conclusion

Few creatures occupy a lower place of moral concern in our society than chickens. Even McMahan, in an account that is otherwise solicitous of the interests of animals, writes that "most of the animals that could be humanely reared for human consumption are connected to varying degrees: pigs more than cows, and cows more than chickens" (2016b: 85; cf. Budolfson 2016a: 171). The new understanding of avian cognition means that claims that chickens are trapped in an eternal present can no longer be taken for granted. Such claims now require empirical evidence. Even if some of the new research is disproven, it should cause us to recognize the standard of proof that must be met in the debate over humane slaughter. This includes not only the anti-speciesist defenses of human slaughter discussed in the previous chapter, but any future attempts

[15] I have been concerned with the ethics of killing a chicken merely to satisfy a culinary preference for meat. But if we would otherwise starve to death, the slow-to-kill principle would offer a more permissive judgement, whether or not we adopted Kymlicka's view of equal consideration, on the grounds that a fully self-conscious entity suffers a greater loss through death than a primitively self-conscious one.

to defend humane slaughter that will invoke protectionist claims. Approaches that do not appeal to up-to-date research in animal cognition are no longer adequate.

My reading of the empirical literature on avian cognition is that chickens can reasonably be said to possess the moral standing that accompanies type-two self-consciousness, when such moral standing is assigned in a manner that gives the birds the benefit of the doubt. And I have argued that the moral entitlement which best corresponds to this attribute is neither a right to life nor a kill-with-indifference stance, but a slow-to-kill stance. An upshot of this claim is that the most widespread form of meat eating in our society is indefensible. In a world in which billions of birds are killed for culinary purposes, this is moral knowledge worth having.

CHAPTER 7

The Logic of the Larder

Introduction

A long-standing view holds that animals benefit from the fact that we eat them. Virginia Woolf's father, Leslie Stephen, articulated this idea in 1896: "The pig has a stronger interest than any one in the demand for bacon. If all the world were Jewish, there would be no pigs at all. He has to pay for his privileges by an early death; but he makes a good bargain of it" (236). Since Stephen wrote, the same thought, without the anti-Semitic flourish, has often occurred to other writers. R. M. Hare, for example, echoes Stephen when he reflects on the situation of fish in a trout farm near his home. The farm is a benefit to the fish, Hare argues, insofar as it is the reason they exist. "I am fairly certain that, if given the choice, I would prefer the life, all told, of such a fish to that of almost any fish in the wild, and to non-existence" (1999: 240). Stephen, Hare, and many other writers note that, all else being equal, a meat-eating world will have more animals in it than a vegan world. So long as the animals in question are well treated, they suggest, it is better for them to come into existence than not (Hare 1999: 240; Scruton 2000: 100; Schedler 2005: 502–3; Callicott 2016: 59). If so then we should embrace some form of meat eating to advance this outcome.

This argument has been dubbed the Logic of the Larder (Salt 1914). We can call it the *Larder* argument for short. Although it predates other immanent arguments against ethical veganism, it shares with them an appropriation of protectionist claims. In particular, the Larder argument takes as its point of departure a commitment to animal well-being. In recent decades such a view has drawn support from arguments in population ethics that maintain the best state of affairs is one in which well-being is maximized. On such a view, a world with more happy animals is morally superior to one with fewer. Thus there is a prima facie reason to support a food system that includes animals living pleasant lives (those that are free-ranged rather than factory farmed) over a world without such animals.

The Logic of the Larder raises complex issues regarding the moral status of entities that do not exist, but which we might bring into existence. Adequately responding to it requires grappling with the underlying theory of population ethics on which the argument is based, the so-called *total view*. This view holds that that when making moral decisions we should grant weight to the interests of the total number of individuals affected by our actions, including individuals whom our actions might create. Although the total view has attracted extensive commentary, no existing responses seem entirely adequate, hence it is necessary to challenge it in a new way. I do so by seeking to replace it with an alternative theory of population ethics, one based on the following principle:

> A necessary condition of a choice being morally good (i.e. morally obligatory or supererogatory) is that failing to bring it about will be bad for someone.

If this principle is true, then the Logic of the Larder is drained of its force, as the theory of population ethics on which it depends is not cogent. The view I defend emphasizes the importance of avoiding a bad outcome. In particular, it is concerned with avoiding a bad outcome that occurs when we do not perform an action or make a choice. Call this the *avoidance view* for short. The case for such a view is partly motivated by the unattractive nature of the total view and other existing theories of reproductive ethics. I advance the strong thesis that the avoidance view is less counterintuitive than either. If the considerations I adduce fail to establish the strong thesis however they may demonstrate a weaker claim. It is that the avoidance view is only as counterintuitive as the total and prior existence views, and so warrants serious contender status alongside them. Although my explicit focus is on the strong thesis, the weaker one lurks in the background as a possible alternative upshot of my account.

After outlining my view, I note how it offers a new reply to the so-called symmetry argument, which posits symmetrical obligations in reproductive cases involving potentially happy and miserable individuals. The avoidance view concedes that there are symmetrical features of the two cases, but it also picks out a crucial difference that justifies asymmetrical obligations. In addition to evading the symmetrical argument, the avoidance view is also able to withstand three possible objections. The first charge is that it cannot adequately address cases of procreation in which the lives that are created are reduced in some way, either in duration or quality, yet are still worth living. Second, harm avoidance entails accepting part of Derek Parfit's famous Repugnant Conclusion, to be sketched later. Third, the harm avoidance view is contradictory.

After defending my view from objections, I briefly discuss a variation of the Logic of the Larder argument put forward by Jeff McMahan. McMahan's variation asks not whether we are obliged to bring food animals into the world but whether it could one day be permissible to do so. So long as certain conditions are met, I argue, it could be permissible, although as we will see the conditions are currently scientifically impossible and are likely to be difficult to meet even if they do someday become feasible. My main focus however is on making the case for the avoidance view. In outlining and defending my view I acknowledge that it has off-putting features. But these features, real as they are, are not grounds to abandon the view. Rather, they are the price one pays for advancing a population ethic that is consistent and does not resort to ad hoc–ery.[1] The case for my view is thus not that it has no unattractive features. It is rather that it has fewer such features than the alternatives.

Benefit Versus Replaceability

The Logic of the Larder comes in more than one formulation. Much of the debate to date has focused on a version that associates it with the notion of replaceability. This version of the argument has been prominently advanced by Peter Singer (1993: 121–31; 2011: 104–119). As Singer interprets Stephen's remark, it highlights the fact that while meat eaters are responsible for the death of food animals, they are also responsible for their existence. "The loss meat eaters inflict on one animal is thus compensated for by the benefit they confer on others" (2011: 105). Singer cites hedonistic utilitarianism as an example of a moral theory that would lend support to this conclusion, insofar is it supports maximizing overall happiness. So long as food animal are not factory farmed and lead happy lives before slaughter, their existence will increase overall happiness.

Singer's replaceability argument has a broader application than animals. As we saw, it also entails the permissibility of infanticide when the parents deem that an existing infant will have a less happy life than a potential replacement child, and when no one wants to adopt the existing child

[1] As Clark Wolf notes, "there is no normative theory of population choice that does not have seriously counterintuitive implications" (2004: 61).

(2011: 163). The replaceability construal of the Logic of the Larder has been subject of extensive criticism. One objection holds that the Larder argument loses considerable force when the moral status of animals and infants is framed in terms of rights rather than utilitarianism. On a rights-based view, animals and infants have a right to continued existence, which can only be overridden in exception circumstances, as when doing so is the only way to avoid a catastrophic loss of life. The utilitarian argument, however, takes the creation of one life to be enough to justify the loss of one life, a far lower standard. If sentient beings do have rights, then the wrong done to one individual when it is killed unnecessarily is not made up for when another is brought into existence, whether animal or human.[2]

My account is closer to the rights-based view than utilitarianism. Because it takes a form that denies the applicability of the replacement argument, and because the replacement argument has already been subject to powerful criticism, I do not take it up further (Višak 2013: 46–70, McMahan 2002: 346–58). It bears noting however that replaceability is not an essential feature of the Logic of the Larder. We might imagine a single animal being brought into existence. It leads a pleasant life and, rather than being slaughtered, is allowed to live out its natural life. After it dies of natural causes it is then eaten (McMahan 2017). On one construal of Stephen's remarks, no wrong will have occurred, insofar as the animal is never harmed at any point. This construal of the Logic of the Larder has less relevance as a defense of contemporary meat eating, which does involve slaughter. But the questions of animal ethics are not all immediately practical. Some are more purely philosophical. What should we say about this kind of case? It is this question my account focuses on.

The Total and Prior Existence Views

Theories of reproductive ethics are commonly categorized according to how widely they draw the sphere of moral concern. The total view holds that when making moral decisions we should grant weight to the interests of the total number of individuals affected by our actions, including individuals whom our actions might create. As Peter Singer, a proponent of the total view puts it, such a view entails that "it is good to increase the amount of pleasure in the world by increasing the number of pleasant lives, and bad to reduce the amount of pleasure in the world by reducing the number of pleasant lives" (1993: 103). Singer speaks of increasing pleasure,

[2] I own this point to Josh Milburn.

but this is not an essential feature of the total view. One could just as easily refer to maximizing preference satisfaction or well-being, or giving weight to welfare interests. What is important is the underlying commitment to maximizing some good, whether it be pleasure or happiness or anything else. The total view holds that we should grant weight to the total amount of such a good. On such an approach, when we can increase the number of entities leading pleasant existences without negatively affecting others, we should regard it as morally obligatory to do so.[3]

This view has some counterintuitive implications. Imagine a couple deciding whether to have a child. Suppose they knew that their offspring would likely live a pleasant life. As we normally view the ethics of reproduction, a potential child's well-being does not oblige its would-be parents to conceive. Couples are ordinarily thought to commit no wrong when they are able to conceive happy progeny but do not. The total view, however, entails that prospective parents commit a moral wrong when they could bring a happy child into the world but fail to do so. Applied to the ethics of bringing animals into existence, the total view maintains that when the question of bringing them into the world is separated from the question of killing them, a world with more happy animals is morally superior to one without. We will have to wait for the animals to die of natural causes before eating them, but insofar as we bring them into the world and they live happy lives, we create a morally better world.

In contrast to the total view is the prior existence view.[4] It holds that when making decisions involving potential individuals, we should grant no weight to the interest of potential individuals themselves. As proponent David Heyd puts it, "choices about whether to bring possible people into existence can and should be guided exclusively by reference to the interests, welfare, ideals, rights and duties of those making the choice, the generators, the creators or the procreators" (1994: 96). On this second view, we have no obligation to maximize happiness or any other good by bringing more people into the world. This is because, when deciding how to act, we need

[3] My discussion skips over an argument Derek Parfit offers for the total view, the Mere Addition Paradox (1984: 419–41). Part of the paradox involves a comparison between two worlds, A and A+. A+ contains all the people who exist in A along with a second group that does not. Parfit generates a paradox from an argument that characterizes A+ as not worse than A. However, it is consistent with the view that A+ is not worse than A to say that the two worlds are incommensurable. If so, then no paradox arises. For a discussion of incommensurable cross-world comparisons, see Handfield (2011). Handfield is concerned with absent desires but the same considerations apply to cases involving absent people, such as those who are absent in A under Mere Addition.

[4] The name of the prior existence view comes from Singer (1993: 103–4).

not grant weight to the interests of potential individuals whom our decision might bring into existence.

It might seem redundant to refer to "potential individuals whom our decision might bring into existence," rather than potential individuals full stop. The longer phrase, however, is necessary in order to avoid a possible misunderstanding. The possible misunderstanding is to think that on the total view alone we should grant weight to the interests of potential people, but this is not what distinguishes the two theories.

On both accounts, it would be wrong to set a bomb that will go off in two hundred years and kill thousands of people. It would be wrong on the total view because such an act would reduce total well-being. But planting such a bomb would equally be wrong on the prior existence view. For the people in the future will exist whether or not we plant the bomb.[5] Their existence is in that sense independent of, or prior to, our decision regarding the explosive device. Given "prior" existence of this kind, which is conceptual rather than temporal, our deliberations regarding the bomb need to take the interests of future people into account, which rules out such a harmful act.

What really separates the two theories is their application to cases where someone is deciding whether or not to bring an individual into existence. In such instances the existence of the potential individual is not conceptually prior to, but bound up with, the decision at hand. On the total view, whatever other factors the couple takes into account, they need to grant weight to the potential future happiness of the entity they could create. Hence the moral wrong they would commit by not conceiving the happy child. On the prior existence view, by contrast, the couple would commit no wrong by remaining childless, even if they knew their offspring would be happy and not reduce anyone else's happiness. Yet they would similarly do no wrong if they conceived the miserable child. Similarly, on the prior existence view they have no obligation to bring happy animals into existence.

Singer suggests that the prior existence view is more in harmony with our everyday views regarding the ethics of reproduction (1998: 384). I think he is right about that. We do not normally hold that fertile couples have a moral obligation to give birth. Yet as we will shortly see, an argument

[5] Leave aside for the moment unusual scenarios in which someone who exists two centuries from now somehow owes their existence to our decision to plant the bomb. Such cases are a version of the Risky Scenario, discussed later.

presented by Singer lends powerful support to the total view and call some of our most settled convictions regarding reproduction into question.

Before outlining that argument, it bears noting that both the prior existence view and the total view have counterintuitive implications in cases beyond individual conception scenarios. The unattractive features of the prior existence view are highlighted in a scenario presented by Parfit known as the Risky Policy (1984: 371). Suppose we pursue an energy policy that will determine who will exist three centuries from now. Perhaps we make cheap energy so plentiful, for example, that the price of air travel plummets and long-distance relationships become the norm. This affects which children are born, as many relationships would not occur but for the energy policy. Suppose further that the policy has such an impact that hundreds of years in the future there is a community we can call Energyland where everyone owes their existence to the policy: each person has at least one ancestor who owes his or her birth to the policy.

But now suppose that the same energy policy also involves burying nuclear waste that is eventually exposed due to an earthquake. People in Energyland are exposed to the radiation, the effect of which is to cause their offspring to die at age 40, with no other effects. The offspring thus owe both their existence and their premature death to the energy policy.

On the prior existence view, no wrong has occurred in the Risky Policy scenario. This is because the people who experience the tragedy in future also owe their existence to the same policy that causes the tragedy. There was never a possibility they could live and also avoid the tragedy. On the prior existence view, no wrong has been done to them. This conclusion is one many people find difficult to accept. Yet it undeniably follows from endorsing the prior existence view.[6]

The total view has additional problems of its own. It entails Parfit's Repugnant Conclusion (1984: 381–90). Imagine world A, in which the average level of happiness is high. In world B, the average level happiness is slightly lower, but there are more people, so that world B contains more happiness than world A. On the total view, world B is better. But so in turn is world C, which has a still lower average level of happiness, and still more people. This process continues until we come to world Z, with an enormous number of people leading lives just barely worth living. This world is best of all on the total account. Such a view is one many people find counterintuitive. But as is now well known, it follows from embracing the total view.

[6] Singer (1998: 394–5) notes further paradoxical and counterintuitive outcomes of Heyd's approach.

The Avoidance View

The prior existence view and the total view are not the only possible approaches one might adopt to population ethics. They illustrate a common attribute of all such theories in this area, which is that each entails a counterintuitive implication of one kind or another.[7] Given this well-documented aspect of theories of reproduction, it is worth asking whether an alternative theory is available that has fewer such implications. I believe such a theory is available. It can justify an asymmetry in our reproductive obligations by reference to the following principle:

> Avoidance Principle: A necessary condition of a choice being morally good (i.e. obligatory or supererogatory) is that failing to make it will be bad for someone.

This principle compels us to avoid a negative outcome, so let us call it the *avoidance principle*. The correct conclusion to draw from considering the case of the miserable child is that the avoidance principle is true, which explains why we have an obligation not to conceive in the miserable case but no obligation to conceive in the happy case. This is because in considering a choice or course of action one of the questions we should ask is, what would happen if we did not do it?

When not performing the action is bad for someone, the condition outlined by the principle is satisfied. Thus when we consider the choice of non-conception in the miserable case, we ask: what would happen if we did not take that tack, and opted for conception instead? The outcome would be bad for the child. When we consider conception in the case of the happy child however, were we to fail to perform it, the result would not be bad for the potentially happy child, as she would simply not exist. In such an

[7] The average view holds that the best outcome is one with the highest average well-being. Among other problems, this view entails that the best possible world is one in which one extremely happy or satisfied individual exists (Parfit 1984: 420). The saturating-counterpart view holds that the best outcome is one that results in the best aggregate of harms and benefits, with harms and benefits determined by "compar[ing] the welfare of an individual in an outcome with the welfare of that individual's counterpart in the alternative outcome" (Višak, 2016: 123). This view entails that parents who deliberately conceive the miserable child do no wrong. It does hold that "if no better option is available, an act of euthanasia or abortion should then prevent the child's suffering," on the grounds that once the child exists its suffering is morally significant (ibid.). But in cases in which it is not possible to end the lives of children so conceived, "the theory cannot condemn their suffering" (ibid.). The repugnant view embraces the Repugnant Conclusion (Tännsjö 2002). The problem with this view is self-explanatory. Cowie (2017) argues that the lives of people in the Z world are not repugnant if we imagine them as lives we might create. This, however, is not an essential feature of the Repugnant Conclusion. The scenario retains its probative force so long as we imagine the people in Z world not owing their existence to us.

outcome there is no one who experiences anything bad. Similarly, failing to opt for conception in the miserable case is not good, nor is failing to opt for non-conception in the happy case. Of the four possible choices, the only one that meets the condition of moral obligation set by the avoidance principle is the choice of not conceiving the miserable child. All else being equal, conceiving the miserable child will be morally wrong, while both conception and non-conception of the happy child will be morally permissible.

Applied to the issue of bringing animals into the world, the avoidance view ask what the consequence would be of not conceiving them. No wrong is done to the animals, as they never exist. There is thus no animal-related reason to bring them into existence. Their merely potential well-being is no more a reason to bring them into the world than is the potential well-being of a happy child. The Larder argument is thus sapped of strength, insofar as it is now seen to rely on a claim, namely that we are morally obliged to create a world with more happy animals in it, that is not correct.

The avoidance principle seeks to capture something important about what it is that makes an action morally good. This feature is that performing the action prevents something bad from happening to an individual, whether human or animal. Normally when we do something good for someone this condition is met. To see this, consider a decision whether or not to place some existing infant in a state similar to the miserable child. Perhaps we image injecting a one-day-old infant with a chemical that will cause it to live a life of agony for two years before dying. If anything is morally wrong, surely this is. Not injecting the child unambiguously complies with the avoidance principle. Injection is bad for the child, so non-injection meets a condition of being morally good. The same will be true of any obviously wrong moral action. Refraining from killing or stealing is morally good, in part because killing and stealing are bad for people.

I say "in part" because the avoidance principle is a necessary rather than sufficient condition of an action being morally good. We also need a standard of justification to determine whether an action is right or wrong. Killing, for example, will not be murder if it is an act of self-defense or part of a military attack conducted according to just war principles. More controversially, some would say that capital punishment is not murder, or that abortion is. Much of the work of determining whether an action is morally good depends on the standard of justification we employ. It is hard to think of any plausible standard of justification that would endorse injecting the newborn or creating the miserable child. Important to note however is that the avoidance principle does not appeal to this fact. The bad outcome has to do with the fact that the subject of the action has an interest that is in some way frustrated by the action. It is bad for someone to be murdered or killed in self-defense, war, etc. whether or not the action is justified. Similarly, if some evil sadist forced us to choose between injecting painful chemicals into one newborn or one hundred, it would arguable be justified to inject the one. But even though the action was justified, it would still be bad for the child. It is this sense of "bad for" that the avoidance principle refers to, which in the broadest sense refers to interests of the subject that are in some way frustrated or not met. The avoidance principle leaves aside the important work of outlining a full theory of justification and focuses on the more modest goal of noting a necessary condition of an act being good. In order to be so, a failure to perform the action has to be bad for someone.

This is not quite the same as saying a failure to perform the action has to *harm* someone. One might think that both the injected newborn and the miserable child post-conception experience harm. Singer for example holds that the happy child gains from being born and the miserable child is harmed by birth (Singer 1990: 228; see also Parfit 1984: 487–90). Yet whether the happy child benefits and the miserable child is harmed by birth is dubious. Parfit has noted that ascriptions of harm and benefit must meet a two-state requirement: "we benefit or harm someone only if we cause him to be better or worse off than he would otherwise at that time have been" (1984: 487). In the case of the miserable child, however, there is no worse state he would have been in if he were not conceived, for he would simply not have existed, and so not occupied any state at all. As a result, his birth cannot be a harm. The existing child whom we inject with the painful chemical, by contrast, very much is harmed.

The existing newborn who is injected with chemicals and the child born into misery both occupy position that meet the "bad for" requirement contained in the avoidance principle. To see this, however, we need to note that their situations are bad for them in two different ways, which we can think of as involving relative and absolute negativity.

In keeping with the two-state requirement, the situation of the injected child is negative relative to the alternative state in which the pain of injection never occurred. Its pain is a deprivation of the well-being it would have experienced but for the injection. The child born into misery by contrast is never worse off relative to some alternative possible state. Yet the outcome for it seems obviously negative. The non-relative nature of the conceived child's condition warrants the term *absolute negativity*. There is something intrinsically bad about the state the child is born into. Like the case of the injected child, that of the one born into misery is defined by what it lacks. But here the standard of judgment we bring to bear seems to involve a counterfactual appeal to the minimal requirement of a life worth living. The agony the miserable child experiences causes the quality of its life to fall below any reasonable threshold of a life worth living, due to the sheer intensity of its pain and its frustrated interest in a pain-free life.

This brings us back to the avoidance principle. It says that a condition of an action being morally good is that *not* performing the action must be bad for someone. Crucially, an outcome can be bad for someone because it is relatively *or* absolutely negative. Both states meet a sufficient condition of being classified as "bad for" their sufferer. Thus in either case, the condition specified by the avoidance principle is met. When we turn to the case of the child who lives a happy existence, by contrast, the condition set by the avoidance principle is not met. Why? Because when the happy child is not conceived, the outcome is not bad for anyone.

If that is the case, then we are well on our way to explaining why we have an obligation to not conceive a miserable individual, whether human or animal, with no equivalent obligation to conceive a happy one. I say well on our way because the avoidance principle is a necessary rather than sufficient condition of an action being good. In order to say that an action is right or wrong all things considered, we need to look beyond the avoidance principle to a standard of justification. I take it for granted that any minimally plausible standard of justification will condemn conceiving the miserable child, whether it be on consequentialist grounds of decreasing overall happiness, deontological grounds such as acting on a principle an affected party could reasonably reject, virtue ethical grounds such as embodying the vice of callousness, or what have you. To be sure, the avoidance principle is incompatible with the total view, and so is not neutral when it comes to deciding between standards. Nevertheless, my goal is to outline a condition that a wide range of theories of justification must take on rather than decide between them, so I do not seek to demonstrate which standard is ultimately correct. Speaking at the general level, when the avoidance standard is not met, an action is not even a candidate for being morally good. Whether one employs a deontological, virtue ethic or non-total consequentialist approach, there is no obligation to conceive the happy child. On each theory conceiving the miserable child is wrong, in part because the avoidance condition is met, plus an additional consideration that each theory will articulate differently, but which in each case has to do with the fact that someone is negatively affected by what happens.

The avoidance principle does important work in explaining why we have an obligation not to conceive a miserable individual, even if the principle is not quite able to offer a complete explanation. In the case of the happy individual by contrast the principle is sufficient to explain why it is not obligatory. It is because a crucial condition of moral goodness is not met, preventing procreation in the happy case from even being a candidate for a morally good action. Procreation in this instance is neither good nor bad, but permissible.

In addition to being less counterintuitive than the going alternatives, a second consideration supports the avoidance view. It is that the avoidance principle outlines a condition that normally obtains when we commit a morally good action, whether it be obligatory or supererogatory. In cases not involving individuals we might create, it is a condition of such actions that if they do not occur the outcome will be bad for someone. This was illustrated by the example of the already-existing infant injected with painful chemicals. We have an obligation to avoid this and other unnecessary harmful actions because failing to make the right choice would be bad for someone. In this way the avoidance view excavates a condition of moral goodness that has long been embedded in moral thought and practice, but heretofore has not been noticed.

The Symmetry Argument

On the view I defend, there is an asymmetry regarding our obligations in two conception scenarios. In this way my view share a trait with the prior existence view, which also entails asymmetrical obligations in the two scenarios. A hurdle any such view must meet is to explain how such an asymmetry is possible. This challenge is one that proponents of the total view have argued cannot be met. They press this point by means of the so-called symmetry argument.

One possible approach to explaining the asymmetry would be along the following lines: "The possibly happy child we fail to bring into existence does not grieve at the loss of the happy life she might have experienced; whereas the miserable child we bring into existence does experience the misery of her existence" (Singer 1998: 385). Singer questions the adequacy of this response, noting that it focuses on only one option in either case. For someone inclined to reject the prior existence view could just as easily reverse the examples: "The possible miserable child we fail to bring into existence does not rejoice at having been spared the misery she might have experienced; whereas the happy child we bring into existence does experience the happiness of her existence" (ibid.). For this reason, Singer concludes that the proffered response does not adequately explain the asymmetry, and the dilemma for the prior existence and other asymmetrical views still stands.[8]

The total view might be said to have its own difficulty here, as it seems to commit us to the position that couples are obliged to increase global happiness through their reproductive decisions. But it will not suffice for a defender of an asymmetrical view to point to this fact and leave it at that. Something needs to be said for the dilemma the symmetry argument presents to the non-totalist, regardless of what problems it also poses for a totalist.

As we just saw, a possible way of affirming an asymmetry was to note that the happy child we fail to conceive does not grieve the loss of a happy life, while the miserable child who is born does experience misery. Singer replied that this response would not do, as the existence of the respective children can be reversed. On his reformulation, the possibly miserable child does not rejoice in being spared a miserable existence, while the happy child does in fact experience happiness.

Singer's reversal does bring out a positive symmetry between the cases. In both instances, when the choice is for non-conception no one ever exists to be thankful for anything. Singer's criticism, however, seems to suggest that when we note this it amounts to the recognition of a profound symmetry between the two cases, after which it is but a short step to endorsing the total view. It bears noting therefore they my analysis does not deny the presence of symmetries *of any kind* between the two cases.

[8] As Jeff McMahan notes, the power of the symmetry argument is that it challenges us to reconcile the common belief that the expectation that a child would have a happy life is not grounds to conceive, but the expectation that a child would have a miserable existence is grounds *not* to conceive. "The imperative to show that these two beliefs . . . are mutually consistent is one of the major challenges faced by those who wish to defend the central pillars of commonsense morality. I believe that this challenge has not been met, and that there is ample reason to doubt that it can be" (McMahan 2002: 353). Allen Buchanan and three co-authors try to meet this challenge by arguing that bringing the miserable child into existence violates its rights, such as its right not to be harmed, a condition they term wrongful life (Buchanan et al. 2000: 232–42). However, Buchanan et al. do not explain how their theory applies to potentially happy child, and whether it too has rights. They also deliberately avoid addressing the repugnant conclusion, to be discussed, and whether their theory can avoid it. Buchanan et al. have also been criticized for combining so-called person-affecting principles and non-person affecting principles in an ad hoc way (Heyd 1994: 9).

A more radical response to the symmetry argument is offered by David Benatar (2006), who argues that coming into existence is always a serious harm. Hence the title of his book, *Better Never to Have Been: The Harm of Coming Into Existence.* For problems with Benatar's view, see McMahan (2009).

In both cases I acknowledge that non-conception meets the minimal condition of being permissible, whether or not one instance of non-conception is also obligatory (as is in fact the case). And also in both cases, I acknowledge the procreative option is not obligatory, whether or not one instance is also morally wrong (as is in fact the case). But alongside these symmetries there is also an important asymmetry: the avoidance condition is met in one case of non-conception but not the other. In the case of the miserable individual, when our choice is for conception it is bad for someone. Thus the choice of non-conception meets a condition, that of avoiding a negative outcome, which fails to obtain when someone opts not to create a happy individual.

For this reason, on the view I am defending, we sometimes have obligations to individuals we could potentially create and sometimes do not. In this way my theory falls in between the total view and the prior existence view. A necessary condition of an action being good on my account is that, if we fail to perform it, the outcome is bad for someone. We have no obligations to individuals who are not made to occupy a negative state, whether relative or absolute, when we fail to perform an action involving them.[9]

My view holds that our obligations are not as extensive as the total view claims. For the total view holds we can have an obligation to bring new individuals into the world simply because they will be happy. The view I am defending denies this. It therefore draws a narrower sphere or moral obligation than the total view.

But my view is not synonymous with the prior existence view. My account draws a comparatively larger sphere of obligation. My theory of obligation covers everyone covered by the prior existence view, whether they be currently existing individuals or individuals who will exist in the future independently of a decision we take regarding them (such as the people who would be harmed by setting a bomb to go off centuries from now). But my view also conditionally includes individuals who will potentially exist as a result of the decision we are making, such as the miserable child and her animal equivalents. We can have obligations regarding such individuals on my account. But a condition of an action being obligatory in their case is that, should we fail to perform it, the outcome will be bad for them. When the action is such that failure to perform does not have a negative consequence for anyone, it is not obligatory on my account. For my account holds

[9] Note that this principle is consistent with an obligation to better people's situations. Failing to better an existing person's situation renders them worse off than they would be if we helped them.

that an individual needs to exist in order for something bad to happen to them.

On this view what some but not all potential individuals share with actual people is that there are actions involving them that, should we fail to perform those actions, will be bad for them. The miserable potential individual stands in this relation to us while the happy potential individual does not. An individual needs to stand in this relation to us, a relation of being negatively affected by our failure to do something potentially involving them, in order for us to have moral obligations concerning that individual.

Possible Objections

Parfit briefly considers a view similar to mine. It is one of a number of views that endorse the following principle:

> The Person-affecting Restriction: This part of morality, the part concerned with well-being, should be explained entirely in terms of what would be good or bad for those people whom our acts affect (Parfit 1984: 394).

Parfit's principle would be more accurately names the *individual*-affecting restriction. As noted in previous chapters, the term "person" is commonly used to refer to entities that are self-conscious, moral agents, etc. The restriction Parfit identifies, however, could apply to any entity for whom things can go good or bad, which can potentially include dogs, infants and other entities that are not usually considered persons. Terminological considerations aside, the core of the restrictive principle is plain enough. Although not identical to the avoidance principle, it too entails that a condition of something being morally bad is that it be bad for an individual. The notion of "bad for," however, can be interpreted in more than one way. Parfit briefly mentions an interpretation he terms the "Narrow Person-affecting Principle of Beneficence," or the Narrow Principle for short. To understand its notion of "bad for," we can compare two possible outcomes, X and Y.

There are people who will exist in outcome X whom we can term the X-people. On the Narrow Person-affecting Principle, outcome X is worse for people "if the occurrence of X rather than Y would be either worse for, or bad for, the X-people" (Parfit 1984: 395). *Worse for* and *bad for* here correspond to my notions of relative and absolute negativity. The Narrow Principle is not the same as the avoidance principle. Among other differences, Parfit's principle outlines a sufficient condition of something being worse for people, while mine outlines a necessary condition of moral goodness understood to include moral obligation. A crucial element of my principle is that we judge the goodness of an outcome by reference to our failure to bring it about. The Narrow Principle comes with no equivalent element. Once we have the avoidance principle in mind, however, it is possible to read the Narrow Principle as overlapping with it on a key idea. This is the claim that when an outcome is not negative for anyone who ever lives – is neither worse for nor bad for them – that outcome is not morally bad. As Parfit rejects the Narrow Person-affecting Principle, his grounds for doing so could potentially justify rejecting the avoidance principle.

The first of Parfit's criticisms is that the view at hand cannot offer moral grounds on which to reject the risky energy policy. As we saw, this was a policy that creates people with lives worth living but also cuts those lives short. Parfit would also say that views such as mine entail accepting part of the repugnant conclusion. Suppose the Z world somehow comes about. This would not be relatively or absolutely negative for anyone who ever exists. On views such as the one I defend, "we must claim that Z would not be worse than A," which Parfit argues is hard to accept (Parfit 1984: 395). Finally, views such as mine entail contradictions according to Parfit, when we note two possible outcomes:

1. Unlike other existing people, Jack has a life that is much worse than nothing. Jill never exists.
2. Unlike other existing people, Jill has a life that is much worse than nothing. Jack never exists.

Parfit concludes:

> If what happens is (1) rather than (2), this would be bad for Jack, and would not be good for Jill. (1) would therefore be worse than (2) for the (1)-people. By similar reasoning, (2) would be worse than (1) for the (2)-people. The Narrow Principle implies the contradictory conclusion that each of these outcomes would be worse than the other (1984: 395).

For these reasons, Parfit concludes we should reject a core aspect of my proposal, which hinges on the same notion of "bad for" as does the Narrow Principle.

I believe we should affirm the avoidance view despite these criticisms. Regarding the first, it is true that my view cannot give a reason to avoid the Risky Policy. Whether this counts as a reason to reject the theory, however, depends on the standard we use to decide between theories pertaining to the ethics of reproduction. If the standard is that a counterintuitive outcome is grounds for rejection, then every theory pertaining to the status of potential people must be rejected, as all have some seriously counterintuitive implications. The alternative to an individual-affecting theory is one which holds that a situation can be morally better or worse whether or not it is better or worse for any individual who ever exists. The total view is a theory of this kind. As I have indicated, it has hugely counterintuitive implications of its own. The appeal of the avoidance view is thus not that it is entirely congenial to our pre-theoretical intuitions, but that it has fewer challenging features than any alternative that rejects an individual-affecting standard. Speaking for myself, I find it very difficult to reject the thought that we do no wrong when we fail to conceive a child whose life would be worth living. Insofar as alternatives to the avoidance view reach a different conclusion they seem even more difficult to accept, the Risky Policy notwithstanding. Similarly, although my view and the Prior Existence view both permit the Risky Policy, Prior Existence cannot condemn the decision to conceive the miserable child. My view can.

Parfit's second criticism held that on an individual-affecting standard, we cannot prefer world A from the perspective of world Z. I am puzzled as to how this can be viewed as a criticism. From the point of view of the world we actually inhabit, there is an alternative world with a smaller number of different people in it where the average level of happiness is higher than in the real world. We stand in a broadly similar relation to this world as the people in the Z world do to the A world. Yet we do not normally think that it would be a better world. Moreover, any standard that we might use to determine that Z is worse than A will not conclude that A is the best possible world. For there is some alternative world, which we might term A+, in which there are even fewer people with higher average happiness. And behind that world is another and another, until finally we arrive a world with one fantastically happy occupant. It does not seem a weakness of the avoidance view or views that affirm an individual-affecting standard more broadly that they entail that the Z world, just so long as it actually comes about, is not worse than A. Some alternative views hold that Z is better than A and, in some formulations (such as Singer's) suggest that we have a moral obligation to come as near to Z world as possible. The avoidance view affirms neither of these conclusions. It rather gives us no grounds to prefer world A until it actually comes about. This seems a limited shortcoming, if it is a shortcoming at all. How much time do we spend in the actual world concerning ourselves with the fate of people who never exist?

This leaves the criticism that the Narrow principle is contradictory. In identifying the contradiction Parfit relies on the following background principle: "If other things are equal, one of two outcomes would be worse if it would be worse for people" (Parfit 1984: 394). Parfit stipulates that World One is worse for Jack, whose life is worse than nothing, whereas World Two is worse for Jill, who is there made to experience an equally grim fate. Both of these conclusions do indeed follow from Parfit's stipulations. But it does not follow that the two worlds are paradoxically worse than each other. This is because in comparing the two worlds, the "If other things are equal" constraint embedded in the background principle is not met. The constraint would obtain in a comparison of two worlds that were the same but for that fact that Jack only existed in one. When we transition from World One to World Two, however, there is a second difference beyond the non-existence of Jack. Jill now exists. As the background principle is not satisfied, we are free to deem the two worlds equally bad or incommensurable, or draw some other conclusion that does not succumb to paradox. That a given world is worse *for Jack or Jill* does not entail that it is worse than the alternative world in which Jack and Jill's fates are reversed.

Science Fiction Omnivorism

Jeff McMahan has inquired into the ethics of an imaginary practice involving potential animals. His imagined practice is one that is currently scientifically impossible, but may someday become possible. He asks us to imagine a future in which "we could genetically modify certain animals, so that they would inevitably just drop over dead painlessly when they reached full size. So we wouldn't have to do any act of killing" (McMahan 2017). McMahan stipulates that the animals in question would lead good lives. While they were alive they would not be harmed in any way. "[Then] they would just painlessly drop over dead. There would be no act of killing. The farmers would just go out into the fields and collect the bodies on schedule" (ibid.).

McMahan notes that drawing a reliable judgment about this scenario requires us to get clear on the issue of non-identity. The animals in question would not exist, would have no identity, but for the fact that we brought them into the world in order to eat them. Our practice therefore does not deprive them of any good or benefit they might otherwise have had, as the only other possible option in their case was non-existence. "If they get these good lives with the practice, [and] nothing is done to them that's bad or worse for them, it's hard to see how that could be objectionable relative to never coming into existence at all" (2017).

I have argued at length that we have no obligation to bring such animals into the world. This is because, from the standpoint of a world in which such animals do not yet exist, no moral wrong occurs when we fail to bring them into existence. McMahan, however, asks us to answer a question not of obligation but of moral wrong. Would the practice he describes in any way be wrong?

McMahan's discussion has a speculative cast, and my answer is offered in a similar spirit. I can think of several contingent features of the scenario that do raise the possibility of wrong occurring. One concerns the effects of confinement. It might be the case that animals in captivity are not able to exhibit all of their natural behaviors. Similarly, depending on the precise timing of when they are programmed to die, it may have negative follow on effects concerning their relationship with their kin. I assume, though, that the scenario could be revised so such contingencies do not pose a problem. We might for example further genetically modify the animals so their natural behaviors and kin relationships are not frustrated in any way.

A more serious concern arises out of how McMahan sets up the scenario. He envisions it as arising out of future scientific breakthroughs that make it possible to genetically modify animals. But this raises the possibility that at least in the early days of the practice it will involve genetically modifying *already existing* animals so as to shorten their lives. We can imagine for example a trial and error process whereby scientists test their new procedure on the embryo of animals that have already been conceived, seeking to transform their normal genes so as to achieve a foreshortened life.

Insofar as the process McMahan described entails any transition period of this kind it will involve an issue mentioned in our discussion of the doctrine of double effect. As we saw there, part of the rationale for the doctrine is the claim that harmful direct agency involves a particular sort of wrong. To recall Warren Quinn's words, the agent of such harm sees his victims "as material to be strategically shaped or framed by his agency . . . [the victims are] usefully involved in the promotion of a goal only at the cost of something protected by their independent moral rights (such as their life, their bodily integrity, or their freedom)" (1989: 349). Insofar as already existing animals are genetically modified to make them live shorter lives, this will involve a double wrong: there is the immoral consequence of the loss of life, and the immoral attitude that sees them as objects to be used.

To avoid focusing on this problem, we can imagine that an evil genius develops the ability to create modified animals but rather than share knowledge of how to create them, simply presents us with the option of creating them or not. We would thus be responsible for their existence, but not for the fact that some could have otherwise lived longer lives. Clearly at this point we are far beyond a defense of any existing form of omnivorism. The scenario however is meant to draw our attention to the underlying link between eating an animal and bringing it into existence. Would eating it be so bad if that was also the reason for its existence?

It is helpful here to separate the question of whether it would be impermissible to create a happy animal with an artificially shortened life from the separate question of whether it would be impermissible to eat it. We have stipulated that the animal in question will lead a worthwhile life while it is in fact alive. Given this, no harm is caused to it by being caused to exist. On the view I have presented here, it is permissible to create such an individual, as they are not deprived of anything, and so not harmed in any way, when we bring them into existence.

To assess the question of eating them, however, we need to know whether doing so is motivated by speciesism. After all, we can just as easily imagine bringing into existence members of our own species who are genetically modified to lead shortened lives. They too simply keel over dead at some point. McMahan gives the example of human beings genetically designed to die at the age of 30 or 35, after which their organs can be used to save the lives of people needing transplants. Why should that not be permissible? McMahan suggests, plausibly, that offering a conclusive reason for condemning the creation of such human beings requires a solution to the non-identity problem. As it is, however, there is no clear reason why it would be wrong to bring into existence a human being who would live an artificially shortened but otherwise decent life. Or to be more precise, there is no reason involving a harm or deprivation to the created human. Until the non-identity problem gives way, any reason not to do so will have to appeal to side effects on third parties. Perhaps some other people will come to love the created humans, and be devastated by their early deaths. Perhaps human life in general will be cheapened and not held in the proper respect, in a manner that has negative effects for others.

In the case of animals created with deliberately shortened but otherwise worthwhile lives, there is another issue to take into account. As we will see in Chapter 9, edibility is a political concept. Traditionally the view that animals are edible but no human beings are has been based in speciesism.

Non-human lives have been viewed as worth far less than those of human beings with otherwise similar cognitive capabilities. Insofar as the science-fiction omnivorism McMahan asks us to consider focuses on only non-humans, we need to imagine it being done in a manner that does not rely in any way on speciesism. This might take the form of bringing into existence human beings with cognitive capabilities at the same level as animals, if not for the purpose of eating them, then for some other purpose which is hastened by their artificially early, but not harmful, death. Alternatively we can imagine human persons with normal cognitive capabilities brought into the world for a similar reason, as in McMahan's example of thirtysomethings whose early-onset deaths enable organ transplants. It is hard to see how a genuinely non-speciesist version of science fiction omnivorism could be compatible with a blanket ban on creating human beings who lead shortened but otherwise decent lives. Alternatively, insofar as science fiction omnivorism was motivated by speciesism, this would be grounds to reject it on a protectionist account. In the matter of creating living beings with deliberately shortened but worthwhile lives, we come to an area in which it will be more, rather than less, difficult to justify inflicting the practice on an animal rather than a human. The need to ensure the practice avoids speciesism provides an additional justificatory challenge in the animal case not present in the human one.

To say the practice faces challenges, however, is not the same as saying it is unjustified. To sum up our discussion of science fiction omnivorism, it could potentially be justified if the following conditions are met: the animals lead decent lives, implementing the practice does not involve shortening the lives of any already existing animals, there are no negative effects on third parties, whether animal or human, and the justification for classifying animals as edible does not presuppose speciesism. Were all these conditions met, it seems there would be no moral barriers to eating such animals. Taken together however these justificatory hurdles seem daunting. Particularly in an age when in vitro meat is an increasingly real possibility, it is not clear why anyone would want to pursue such a path. But our concerns in this chapter have been primarily philosophical. And however unlikely the scenario at hand may be, on a philosophical level, it seems appropriate to acknowledge that we here have conceived a form of new omnivorism that would be ethically permissible.

Conclusion

I have defended the asymmetry in our everyday view regarding the ethics of reproduction by arguing that a necessary condition of a choice being morally good is that failing to make it will be bad for someone. When this principle is combined with any of a number of different theories of justification, it yields the conclusion that we have a duty not to conceive the miserable child and no equivalent duty to conceive the happy child. In this way, the avoidance view is able to withstand both the symmetry argument and Parfit's criticisms of the Narrow Person-affecting Principle of Beneficence.

A negative feature of the view I have been defending concerns the Risky Policy.[10] There is no grounds on the avoidance account to say that it is wrong, given that the people the policy creates lead minimally decent lives, and thus are not placed in a category of absolute negativity. Similarly, the people who are ultimately killed by the Risky Policy also owe their lives to it, and so we cannot say they are harmed by the policy. In short the Risky Policy is neither bad for not worse for the people of Energyland. When individual-affecting views such as mine are examined in isolation, this stands out as an off-putting feature. However, it seems unlikely that a theory will emerge that will leave intact all of our common-sense views. Thus the real test of a theory of population ethics is how it compares to other theories on the subject. By that standard the avoidance view is, to paraphrase Churchill, the worst possible theory except for all the others.

[10] My theory also obliges us to accept the implications of other thought experiments Parfit presents, including Depletion (1984: 362). As the Risky Policy strikes me as the most counterintuitive, I have focused on it. If it can be stomached, Depletion and the rest are comparatively easier to accept.

CHAPTER 8

Thinking Like a Plant

Introduction

Animal protection arose during a historical moment, the 1970s, when factory farming defined the treatment of food animals. Agriculture, however, is a human invention that changes over time. The rise of humane slaughter, examined in Chapter 5, is but one example of how modern agriculture has changed relatively recently. As I noted there, one reason behind the rise of human slaughter was the work of animal protection advocates such as Peter Singer and his former student Henry Spira, both of whom played a role in causing the McDonald's corporation to adopt humane slaughter. However, new values are not the only drivers of institutional change: scientific innovation can also play a role. And it is conceivable that science might foster breakthroughs in our understanding that make omnivorism safe for animal protection. Indeed, recent findings in the plant sciences are now said to have already done so.[1]

Contemporary botany has witnessed an upheaval in its understanding of the electrophysiology, cell biology and signaling systems of plants. An insurgent school of botanists have coined the phrase *plant neurobiology* to describe a new field that is "aimed at understanding how plants perceive their circumstances and respond to environmental input in an integrated fashion" (Brenner et al. 2006: 413). As plants do not have neurons, *plant neurobiology* is a deliberately radical concept. Its proponents credit plants not only with powers of perception but also intelligence, learning, and memory. Although controversial and disputed within botany, neurobiologists have achieved enough influence to inspire some moral philosophers to argue for a revised view of the moral status of plants.

[1] Plant neurobiology and plant thinking have been subjects of considerable public interest (Smith 2016; Pollan 2013; The Scientist 2017; Marder 2012a, 2012b; Cook 2012).

Different philosophical proponents of plant intelligence draw different conclusions about plants' moral status. But simply by questioning the chasm of moral inferiority that has long been thought to separate plants from animals, philosophers inspired by plant neurobiology are predictably viewed as providing a justification for meat-eating. As one such philosopher puts it, his work on the moral standing of plants makes "my non-vegetarian acquaintances rejoice in the idea that those who refuse to eat meat might not be righteous after all" (Marder 2013: 185). Commentators on the rise of plant neurobiology have similarly taken it to "hit hard at the foundation of veganism" (Aloi 2011: 93).[2]

Arguments from plant neurobiology potentially justify some form of omnivorism by broadening the application of a concept employed by protectionism, sentience, into the plant kingdom. Given that we must eat to survive, what could be the rationale for eating sentient plants but not sentient animals? Plant neurobiology is effective in showing that the traditional image of plants as "inert, vacant, raw materials" is outdated (Hall 2011: 156). But even if this did show that plants possess some limited degree of moral standing, this would at most permit some hunting of wild animals. It would still be impermissible to raise animals for food. As it stands, however, arguments from plant neurobiology to date have failed to establish that plants do in fact warrant any degree of moral status. Such arguments therefore do not represent a challenge to traditional veganism.

Plant Thinking

Michael Marder uses the term *plant thinking* to describe the project of attributing perception and other attributes to plants (2013). Let us use the term *plant thinkers* to characterize philosophical proponents of this view. A common view they argue for is that plants possess some form of sentience:

> It is possible to infer plant sentience from the fact that plants explore and pursue unevenly distributed resource gradients, assess environmental dangers from biotic and abiotic stressors and gather and constantly update various types of information about their surroundings (Marder 2012c: 1368).

[2] See also the title of an article by Marder: "If peas can talk, should we eat them?" (Marder 2012a).

The intellectual violence of backgrounding plants and denying their sentience can be said to underpin the "occupation, appropriation, and commodification" of the plant kingdom (Hall 2011: 7).

Expansionary sentientists account for the ample scientific evidence that plants are sentient (Smith 2016: 39).

Emerging scientific research has revealed that plants show all the traditional indicators of sentience. This casts doubt upon the utility of the traditional rigid division made between plants and animals (Pelizzon and Gagliano 2015: 5).

In arguing for the existence of sentience beyond the animal kingdom, plant thinkers employ a particular understanding of intelligence. That understanding is one that they take over from neurobiologists, who in turn took it over from philosopher-psychologist David Stenhouse. Stenhouse defined intelligence as "adaptively variable behavior within the lifetime of the individual" (Stenhouse 1974, quoted in Hall 2011: 143). This definition, which is widely cited by botanists and plant thinkers, is noteworthy for categorizing intelligence as a type of behavior (Trewavas 2007; 231; Hall 2011: 143; Smith 2016: 23). Insofar as plants can exhibit behaviors that are adaptive in the evolutionary sense of advancing their fitness, they meet the criterion of being intelligent. Plants may not be mobile but they are capable of growth and development, activities that contribute directly to a plant's survival and reproduction.[3]

Matthew Hall notes that the ability of a plant to change its outward form has long been recognized by botanists. He cites the example of a member of the *Ranunculus* genus that under water grows multifid leaves, which are leaves divided into lobes by clefts and which have capillary segments. When the same plant grows leaves above water they are larger and broader and do not exhibit multifid division. Other plant species develop in a similarly varied way. According to Hall, this demonstrates that plants are not automatons capable only of preprogrammed behavior. "Plants accomplish this plastic development in a way that by necessity involves assessment of the prevailing environmental conditions and the selection of appropriate responses," he writes (144).

[3] On this conception of intelligence, it will also be possessed by bacteria, viruses and the wide range of cellular organisms that inhabit our bodies. I am grateful to Matthew Fulkerson for drawing this implication to my attention.

Plant thinkers similarly argue that plants are capable of learning. Examples they cite include tendril growth and responses to water stress. It has long been known that tendrils will move in the direction of objects that can support them. If the support object's location is moved, the tendrils will change the direction of their growth in order to relocate it. In a similar way, some plants respond to drought conditions by shedding their leaves. In order for this to occur the plant must be able to determine the present amount of water in its environment, which plant thinkers characterize as a form of assessment.

In this regard Smith cites an experiment that investigated the learning ability of *Mimosa pudica*, also known as the sensitive plant or touch-me-not, the leaves of which fold inward or droop when touched. Experimenters followed a protocol used to test learning in animals by repeatedly dropping mimosa plants 15 centimeters, a harmless distance. They wanted to determine whether the plants, which initially withdrew their leaves upon impact, would come to ignore the harmless stimulus. The experimenters found that the plants did, sometimes reopening their leaves after as few as four to six drops. The possibility that the plants were simply fatigued was ruled out by the fact that when they were shaken they again folded their leaves. The researchers took their findings to demonstrate "the acquisition and expression of a long-lasting memory for a learned behavior in this plant" (Gagliano et al. 2014: 69).

The notion of plant intelligence and learning might be thought to be ruled out by the fact that plants do not have brains. Hall, anticipating this objection, cites recent arguments by plant neurobiologists to the effect that plants may possess "thousands of brain-like entities that are involved in the emergence of intelligent behavior" (147). The entities in question are meristems, a type of tissue located at the tip of shoots and roots. Meristems are connected by vascular strands that are capable of transmitting molecular and electronic signaling. As such, Hall speculates, they could represent a plant's equivalent of the nervous and neuronal systems. One way they may do so is by transmitting the hormone auxin and other signaling molecules. Auxin is created at the tips of roots and shoots, and is thought to play a signaling role within plants:

> Although the exact processes have yet to be uncovered, it has been proposed that this extracellular transport of auxin "exerts rapid electrical responses" across the plant synapse and "initiates the electrical responses of plant cells." Whatever the pathway within the plant, communication can occur over long-distances, with information on the environmental and developmental state of the roots being transferred to the shoots . . . . As well as auxin and

> electrical signals, plants produce and use a variety of neurotransmitter molecules to communicate from cell to cell. Dopamine, acetylcholine, glutamate, histamine, and glycine are all touted as potential signalling chemicals between cells. Other complex communication molecules include protein kinases, minerals, lipids, sugars, gases, and nucleic acids. Trewavas has drawn attention to this complexity and notes that "from the current rate of progress, it looks as though communication is likely to be as complex as that within a [animal] brain (Hall 2011: 147).

Insofar as hormones and electrical signals can transmit information at the same degree of complexity as do brains, Hall suggests, then brains may not be required for the possession of intelligence and learning.

Plant sentientists offer a second reason for taking seriously the possibility of plant intelligence without plant brains. They endorse a systems-based view of intelligence, according to which intelligence, reasoning and choice do not require a unitary mind, understood as something external to or separate from the biological system that houses it (Marder 2012c: 1370; Smith 2016: 24). Intelligence and related concepts are rather the product of decentralized self-organizing networks in all living things. Such a view is summarized by Fritjof Capra:

> According to the systems view, the essential properties of an organism, or living system, are properties of the whole, which none of the parts have. They arise from the interactions and relationships between the parts. These properties are destroyed when the system is dissected, either physically or theoretically, into isolated elements (quoted in Hall 2011: 149).

Hall argues that once we reject the "Cartesian view" of mind that presupposes a mind's location in an isolated organ such as the brain, there still remain questions about the precise systems within plants that give rise to intelligence.[4] Nevertheless, on a plant sentientist account, the lack of a brain ceases to be a barrier to ascribing plants with intellect.

Criticisms of Plant Thinking

Part of the plant-thinking project is to challenge the view of plants as inert forms of life. The findings regarding internal plant communication they

[4] Hall's account may overlook that contemporary philosophical accounts of agency and intelligence also posit a decentralized or fragmented view of the mind. I am grateful to Matthew Fulkerson for this point.

cite suggest that aspects of the internal working of plants are more complex than botany has traditionally allowed. As one critic of plant neurobiology has put it, "current scientific knowledge does not provide a conceptual framework to satisfactorily describe the organization of the complex metabolism, the integration of all signals perceived and sent" by plants (Struik et al. 2008: 364). It is entirely possible that we may need to revise our understanding of the internal and external workings of plants in light of findings cited by plant thinkers.

But is the most appropriate revision one that ascribes to plants intelligence, let alone sentience? It may be more appropriate to revise the traditional image of plants in a manner that stops short of ascribing to them either attribute. Instead of categorizing plants according to concepts derived from our understanding of animals, human and otherwise, it may well be more justified to view them as occupying their own unique category, one that acknowledges their vitality and dynamism, yet stops short of bestowing them with intelligence or moral standing.

Grounds for this more moderate view are provided by plant scientists who do not view plant neurobiology as helping advance our understanding of plant metabolisms. In 2007, 36 scientists and 33 research institutions released a public letter characterizing plant neurobiology as being based on "superficial analogies and questionable extrapolations" (Alpi et al. 2007: 136). One such extrapolation concerns the claim that plant signaling is as complex as the kinds of communication enabled by animal neural systems. According to mainstream botany and neuroscience, even if the most revisionary accounts of plant signaling prove correct, the communication systems of animals will still be exponentially faster and more sophisticated.

Consider the role plant thinkers ascribe to the auxin hormone. Proponents of plant neurobiology equate cells that help transport auxin with chains of neurons. A difficulty with this parallel is that neurotransmitters do not travel from cell to cell over long internal distances the way auxin does. In addition, the public letter argues that neurobiologists misidentify the particular pathway within cells along which auxin travels, falsely saying it occurs in a bubble-like membrane known as the vesicle (Alpi et al. 2007 136). Insofar as auxin does not travel in this way, its function and transportation are still less similar to that of neurotransmitters.[5]

[5] Hall cites plant neurobiologists who respond to this criticism by arguing that "too little is known regarding plant signaling . . . to exclude a role of cell-to-cell transport of a 'neurotransmitter-like' compound" (Brenner et al. 2007: 286; Hall 2011: 148). But lack of knowledge does not seem sufficient grounds to endorse the strong claim that auxin or other plant hormones function akin to neurotransmitters.

There is a further disanalogy between signaling hormones and neurotransmitters. The latter contribute to the delivery of much more complex information. While the molecules of signaling hormones are indeed complex, this is not because they carry complex content, as do the transmission systems of which neurotransmitters are a key component. It is because they need to ensure that the message they contain reaches the correct destination within the plant. In botanist Richard Firn's words, "a signalling molecule is akin to a postcard bearing a tick or a cross as the message, with the much more complex information of the address ensuring that the simple message reaches the right recipient" (2004: 348; cf. Struik et al. 2008). The communication complexity that neurotransmitters contribute to, by contrast, is primarily due to the complexity of the information they help deliver. This is sufficient grounds to doubt that the internal transmission of hormones and electrical signals can perform the same complex communication function as a brain.

There are also grounds to doubt that the response of mimosa plants to being dropped provides evidence of their ability to learn. A more likely explanation is that the action is the result of so-called mechanosensitivity. Plants often receive external stimuli that vary widely in their frequency. Wind, for example, can gust repeatedly in a matter of seconds or not at all for days. "If every response to each mechanical stimulus was of the same magnitude, plants would invest greatly in withstanding mechanical perturbation to the detriment of growth," write one group of mechanosensitivity researchers. "Plants thus need to permanently fine-tune their response to mechanical stimulation in order to avoid the cost of [constant activation of stress responses]" (Leblanc-Fournier et al. 2014: 401). Stress can leave an imprint in a plant that causes it to reduce its mechanosensitive responsiveness to external stimuli. Such variation can have a genetic or biochemical cause (401). Therefore, mimosa experiments do not establish that plants have the ability to learn in any sense of the term that presupposes the existence of a mind.

The criticisms that plant neurobiology has received is a sign of its disputed status among botanists, a status that has been the subject of attention in the media (Pollan 2013). In this way, plant neurobiology has achieved a different level of scientific acceptance than the recent findings regarding the mental abilities of birds we examined in Chapter 6. The scientific changes described there included the Avian Cognition Nomenclature Consortium, which has not met with controversy the way the concept of plant neurobiology has. In addition, our discussion of primitive self-consciousness in animals was confined to one species. Plant neurobiology is widely rejected in part because it makes sweeping claims about a large number of plant species. For these reasons we can in good conscience endorse primitive self-consciousness in chickens while expressing skepticism about plant neurobiology.

However, grounds for doubting the claims of plant thinkers are not merely scientific. They are also philosophical. In particular, there are philosophical reasons to question the notion of intelligence that plant thinkers employ.

As we saw, plant thinkers work with a behavior-based view of intelligence put forward by Stenhouse in the 1970s. Behaviorist explanations of organisms have come under strong criticism in the decades since Stenhouse wrote. Ned Block, for example, has argued that whether behavior is intelligent depends on the nature of the "internal information processing" that causes it (Block 1981: 5). Block, like John Searle and other critics of behaviorism, argues that it is possible to imagine an entity that behaves *as if* it were intelligent, but in fact is not (Searle 1980). An example of a system with behavior typical of intelligent beings would be a robot programmed to behave the way human beings do. As its actions reflect the intelligence of its programmers, it would not itself possess intelligence. The question of the nature of intelligence is, of course, complex. But the fact that it is possible to imagine an entity that exhibits behavior typical of an intelligent being but which is not itself intelligent is enough to show that Stenhouse's attempt to equate intelligence with behavior is questionable. It is a shared flaw of plant thinkers and radical botanists that they assume, rather than argue for, a widely disputed view of intelligence.

We intuitively associate intelligence with consciousness. Thomas Nagel has famously characterized consciousness as something a creature experiences: "an organism has conscious mental states if and only if there is something that it is like to be that organism—something it is like for the organism" (Nagel 1974: 436). The behaviorist view seeks to define intelligence it a way that does not refer to this phenomenological aspect of consciousness. This would seem to miss the central aspect of consciousness. As George Sher puts it, the fact of consciousness is the fact of our "subjectivity or interiority—that is, precisely the fact that each has a private reality that is accessible to him alone," a private reality that "when it disappears, will entirely extinguish his world" (2014: 86). It is this property of animals, and the related capacities of being able to experience pleasure and pain, which have traditionally been denoted by the term "sentience."

When plant sentientists refer to sentience, they would appear to denote something else. Marder, for example, writes that plant intelligence has a decentralized structure. For this reason,

> it does not fall under the category of organismic life, succinctly expressed by French philosopher and physician, Canguilhem: "To live is to radiate; it is to organize the milieu from and around a center of reference, which cannot itself be referred to without losing its original meaning." The plant does not organize its milieu (2012c: 1370).

To organize one's milieu is to exhibit intentionality in the philosophical sense of the term. It is to have mental states that are directed toward some object or state of affair—mental states that are *directed at* or *about* something. Our center of reference in this regard is our consciousness, which we cannot reduce to just another object of perception without losing sight of its unique power to make possible perception itself. Marder is saying plants are not conscious in this basic and familiar sense. Hall puts it more succinctly, without the metaphor of radiance. "The idea of plant intelligence ha[s] little to do with notions of consciousness" (2011: 145). Plant sentience thus amounts to sentience without consciousness.

J. L. Austin famously noted that a vice of philosophical writings was promising conclusions that they do not quite deliver. As he put it, "there's the bit where you say it, and the bit where you take it back" (1962: 2). Plant thinkers say plants are sentient at the level of terminology, but take it back at the level of definition. Once their terminology is clarified, the view they defend eliminates an essential feature of our concept of sentience. This is a fateful omission.

The reason we have been concerned with plant thinkers is because their arguments promise an immanent critique of animal protection. Such a critique works with the major philosophical premises of animal protection and, through appeal to new empirical information or mid-range principles, generates a new recommendation concerning our behavior. Sentience, however, is no minor concept in animal protection theory. It is a central property to which protection theories, despite their many differences, typically ascribe fundamental importance. Plant thinking, rather than exploit protectionism's understanding of sentience, rejects it. As a result it is not really an immanent critique after all, despite how it is often interpreted.

In response a plant thinker might say, "to hell with protectionism's understanding of sentience: the behaviorist view is correct!" This reply however faces a familiar problem. In affirming an understanding of sentience that takes it to involve inner mental states, protectionism makes one of its least controversial claims. The view that it is not so easy to dispense with a phenomenological understanding of consciousness is widely shared. As Sydney Shoemaker has put it, "the philosophy of mind without the first-person perspective, or the first person point of view, is like *Hamlet* without the Prince of Denmark" (1994: 7). Defenders of behaviorist frameworks that seek to explain consciousness without reference to mental states have the thankless task of rebutting the influential criticisms of Block, Nagel, Searle, Shoemaker and others (Dennett 1978). In my view, defending a behaviorist view of intelligence in the twenty-first century is a bit like defending bell bottoms or smoking on airplanes. It may have been acceptable in the 1970s, but our understanding has moved on. Even if I am wrong, the onus is on defenders of behaviorism to defend it in the wake of contemporary critiques.

But perhaps I am wrong, and it is possible for plant thinking to meet this challenge. In order to pose a threat to traditional veganism, however, it needs to be the case that plants occupy a broadly equivalent moral standing to animals. This view however will requires a theory of moral standing and what mental characteristics it is based on.[6] Importantly, most versions of animal protection theory, certainly the most influential ones, have long distinguished between degrees of moral standing. This is true not only of the view I have defended but a wide range of views going back to *Animal Liberation*. On these views if we were in a lifeboat situation in which we had to choose between a dog and a typical human being, it would be morally right to save the human. This is because animal protection theories

[6] There is a strand of the philosophical literature on moral psychology and free will that denies a central role for consciousness (Wegner 2002; Doris 2015; Carruthers 2013). As summarized by Matthew Fulkerson, who drew this this family of theories to my attention, they tend to support the view that "consciousness is like a self-reporting mechanism, a way of broadcasting actions and intentions to the whole system. It is a kind of epiphenomenal narrator that plays the role of agent, but which has no actual causal powers or decision-making capacity" (personal communication, 2017). This deflationary view of consciousness might be thought to show that human and animal minds are far more plant-like than we have traditionally thought, thereby providing a non-behaviorist road to the plant thinkers' conclusions. The deflationary theories in question are comparatively recent and subjects of ongoing debate and criticism that I cannot do justice to here (see Nahmias 2002, 2014; Byrne 2012; Bermúdez 2013; May 2018). In addition, even if deflationary accounts of mind do prove true, they might still afford grounds on which to distinguish animal cognition from plant signaling. It might for example be a significant limitation of plants that they do not even give rise to the "illusion" of conscious will that, according to deflationists such as Wegner, human cognition gives rise to (and which may be true of some nonhuman animals also). For these reasons it seems best to wait until arguments from mind deflationism to plant thinking are worked out and presented in detail before assessing them.

have long posited that persons warrant a higher degree of moral standing than merely sentient beings. Most animal theories are two-level theories in this way.

If it did turn out to be the case that, contrary to our best arguments and science, plants are conscious, it would not necessarily follow that they occupy the same plane of moral standing as animals. A more appropriate response might be to continue to deny them equivalent standing to animals, on either one of two possible grounds. The first is that they are somehow conscious but not sentient. That is, they have something like a first-person point of view – a first-plant point of view – but cannot feel pain. For this reason, they would possess moral standing, but to a degree less than animals. The same would hold true if they turned out to possess primitive consciousness. As we saw in Chapter 6, there are grounds for attributing primitive self-consciousness to chickens, which is a less complex version of self-consciousness. The correct response to a reconstructed version of plant thinking might be to view plants as possessing primitive consciousness, or a less complex version of consciousness.

An alternative view is to hold that consciousness is consciousness, and such fine-grained distinctions are unwarranted. This view, however, needs to clarify whether pineapples have the same degree of moral standing as persons. If I could only bring a plant or a human being into the lifeboat, is there even a question that I should bring the human?

When we examine plant consciousness theories we find few takers for the view that plants occupy the same degree of moral standing as humans. Hall, for example, does not go so far as to rank plants as equal to human beings. "Killing plants is not the same as killing a human being, but killing plants is still considered to be violent" (2011: 160). This view admits that even if plants are intelligent in some sense of the word, there is a still a difference in moral standing between them and human persons. Hall does not identify the qualities that distinguish human from plant persons, but any minimally plausible approach will give weight to all the familiar attributes that are thought to be morally relevant, such as self-consciousness, the ability to form plans and desires and to experience pleasure and pain. Insofar as many of these attributes have some analog in animals, it will make sense for them to occupy an intermediate status between plants and persons (or for some animal species to warrant full-blown personhood, depending on what the best science says about their cognitive capabilities).

What would follow from plants occupying such a moral status? I have argued that they do not in fact possess any degree of intrinsic moral standing. But even if I am wrong and they do have some limited degree of inherent standing, it still would not support modern animal agriculture. It would at most entail a case for eating wild animals.

Rabbits and other animals of the forest can eat plants without killing them, but they kill some of the plants they consume. If such plants have moral standing even to a limited degree, then there is a case for killing, and by extension eating, the animals responsible for their deaths. Such therapeutic hunting campaigns would need to be constrained so as to not risk damaging overall ecosystem functioning. And if it became possible to control wild animal populations through non-lethal means, such as a contraceptive programs, that would be a morally preferable alternative. But when such killing was conducted within appropriate limits, and when non-lethal alternatives were not available, it would represent a morally defensible, even necessary, form of omnivorism. By the same standard, there would also be grounds for killing human persons in order to save the lives of plants, although the number of plants would need to be extremely large, much larger than would be required to justify killing a merely sentient animal.

It would still be wrong, however, to eat animals specifically raised for food. Hall notes as much in a discussion of the global environmental effects of agriculture, which include substantial contributions to both species extinction and greenhouse gas emissions. The novelty of his analysis is that alongside these familiar problems, agriculture harms individual plants that are cleared away to make room for farmland. Despite this eccentric aspect of his analysis, Hall draws a familiar conclusion by singling out animal agriculture as especially damaging. "The rearing of livestock on such large scales is one of the major drivers of habitat loss," he writes. "Basing diets on meat consumption excessively inflates the area of land that is put under human cultivation. Reducing the amount of consumed meat is a direct way of reducing harm done to plants, animals, and human beings" (165). Even if plants occupy a position of moral standing, in other words, a diet that excludes domesticated animals will still be less harmful than conventional omnivorism.

Beyond the sheer amount of land cleared for animal agriculture, there is the further fact that food animals eat plants such as corn, grain, and grass. If these plants are intelligent, they are harmed when food animals eat them. Insofar as we are concerned not to see individual sentient plants killed, it will make the most sense to cut out the middle animal and consume plants directly.[7] Thus even on an interpretation of plant thinking other than the

[7] Smith responds to this criticism by arguing that an agricultural regime of "livestock permaculture" will in some climates be superior to one of "vegan permaculture" (104). Smith relies on Fairlie (2010),

one I have defended, one that overcomes its dependence on behaviorism and grants some moral standing to plants, it will still pose no threat to the prohibition on eating animals save some limited number that are hunted.

Smith, unlike other plant thinkers, thinks otherwise. On his view recognizing plant sentience entails the permissible of consuming domesticated animals. He offers two arguments for this view. The first involves an endorsement of animism. Although inspired by indigenous and other spiritual traditions, Smith's conception of animism is not supernatural. Rather, he characterizes it as the view that personhood is widely distributed in nature. "People can act intentionally and autonomously. They have minds of their own and they use them. These are all characteristics that both plants and animals exhibit" (2016: 53). On this view, human beings do not occupy a position of mastery over the rest of nature. Rather as animals ourselves we are embedded in a larger ecological system in which some persons must consume others to survive. "There is no outside nature, no ontological position of privilege, no ecological separation between humans and other-than-human beings" (46).

Smith urges respect for non-human persons and counsels that it is a mistake to view them *only* as food. But our relationship within a shared ecology is a relationship that requires persons to be mutually available to each other as food. "One can be ontologized as edible – one can be viewed as food – according to ecological animalists without being treated as inferior" (46). We should not subject animals to factory farming and not subject plants to conventional agricultural practices such as heavy antibiotic exposure or monocropping, which can respectively degrade soil integrity and place plants at greater risk of infection (68). If we abide by these restrictions, we can eat both types of persons in a manner that is respectful.

Smith's second argument challenges the very possibility of veganism as a distinct dietary practice. Smith characterizes eating as a transitive activity. If I eat an entity then I simultaneously consume other entities that it has eaten. We are not just what we eat, but what our food eats (or as Smith puts it, *who* our food eats). "Plants acquire their nutrients from the soil, which is composed, among other things, of decayed plant and animal remains. Yes, plants feed on animals" (72). As a result of this, veganism is not merely misguided as a moral doctrine. It also is biologically impossible, as even the staunchest vegan will sustain herself by transitively consuming animals.

whose supporting reasons for this view include the fact that farms that contain animals can make use of manure and other benefits. Smith overlooks the possibility of a farming system that includes animals, and so enjoys such benefits, without killing or mistreating them.

"Vegetarianism is morally indefensible and even ontologically illusory" (122).

Both of Smith's arguments should be rejected. Vegans have always known that plants sustain themselves on the decayed remnants of animals. Avoiding this has never been part of what is meant by veganism. Rather than seek to avoid eating animals "transitively," vegans have sought to avoid eating them *directly*. A primary reason is because doing so normally involves supporting systems that kill animals, and often inflicts suffering on them, in order to provide food to humans. Smith's argument that plant-based diets are ontologically illusory targets a strawdoll version of veganism.

Smith's ontological argument would also seem to open the door to widespread cannibalism. Generally we do not accept that human persons can be "ontologized as edible," but Smith's argument would seem to question the basis of this taboo. Indeed, insofar as eating is transitive, and plants nourish themselves on the decayed remains of human beings as well as animals, by Smith's standard it is ontologically illusory to be a non-cannibal.

Smith is frequently at pains to note he is not defending conventional omnivorism, which permits factory farming and other agricultural practices of which he disapproves. But he fails to note just how far the dietary program his argument entails differs from the everyday understanding of vegans and omnivores alike. Insofar as Smith's criticism of veganism calls into question any distinction between veganism, omnivorism, and cannibalism it is distinctive only for its implausibility.

And yet this is not the worst problem. On Smith's view being a person is no barrier to being food. There is no "ontological position of privilege" that separates us from the many persons he finds when he looks out at agriculture. This would seem to open the door to sacrificing human persons in order to benefit non-human persons. Our normal view that human persons are morally superior to plants, however, is based on the view that human beings possess self-consciousness that distinguishes them from flora. To be sure, Smith states that he rejects a "reductive instrumentalism whereby we can do to and with individuals what we will for the sake of the greater good" (89). Yet if I were driving my car late at night and could only avoid hitting a child by swerving into some bushes, on Smith's account this would be a choice between killing one or several persons. As such, there appears to be no special reason to try to avoid hitting the child. Similarly, there would appear to be no particular reason to rescue a human being if their death would help sustain an ecosystem. The term *environmental fascism* was coined for such views, which have a long and discouraging history in environmental thought (Regan 2004: 362). Smith's theory would seem but another example.

CHAPTER 9

Long Live the New Flesh

Introduction

In 2013 researchers at a lab in the Netherlands created a hamburger by taking stem cells from a piece of beef and exposing it to fetal bovine serum to induce growth. The result was a new kind of hamburger: one made from beef, but which was not carved out of the body of an animal (Fountain 2013). The question raised by in vitro meat is whether it is acceptable from a protectionist point of view. If so, should the resulting food product even be classified as meat?

Like plant neurobiology, in vitro meat is a product of twenty-first-century science. It extends the concept of meat to include flesh that was not derived from the carcass of an animal. Could there really be protectionist arguments against eating meat grown in a lab? In this way in vitro meat also redeploys a familiar concept so as to potentially justify a new omnivorism. Yet despite their similarities, the issues raised by plant neurobiology and in vitro meat differ on a bedrock level. Arguments from plant neurobiology have been interpreted as posing an ethical challenge to plant-based diets. In vitro meat by contrast represents a form of meat-eating that protectionism should condone, even celebrate. Its arrival should prompt us to widen our concept of what meat is to include lab-grown beef, pork, chicken, and fish. It should also oblige us to examine the criticisms of in vitro meat that have been made by thoughtful proponents of animal protection, and show how such criticisms can be overcome. Crucial to doing so is revising our concept of what meat is and can be.

In Vitro Meat

Winston Churchill foresaw something like in vitro meat when he predicted a future in which "we shall escape the absurdity of growing a whole chicken in order to eat the breast or wing, by growing these parts separately under a suitable medium" (1931: 1). Researchers at different labs achieved intermittent success creating small samples of lab-grown meat that mimicked the flesh of species such as goldfish and frogs in the years prior to the 2013 unveiling of the Netherlands burger (Jacobs 2015; Pincock 2007). None of the earlier efforts however achieved the same prominence as the Netherlands burger, which sparked discussion of the future of meat. Even since the Netherlands burger was unveiled, in vitro meat, also known as cultured meat or lab-grown meat, has become easier to make. The Netherlands burger famously cost US$325,000, but the researchers who created it (who subsequently began working on a commercial version) have said the price has since dropped to US$12 (Shoemaker 2015).

The changing nature of in vitro meat bears emphasizing. One reason is because current methods for creating cultured meat commonly involve the use of fetal bovine serum, a growth agent derived from animal fetuses. This serum is a by-product of the dairy industry which, due to the constant lactation it imposes on sows, eventually renders them unable to produce milk. Cows at this stage are often too lame to be used productively and so are slaughtered to produce ground beef or pet food. When former dairy cows are pregnant at the point of slaughter their fetuses are removed and have fetal bovine serum extracted. The use of such serum means that in vitro meat in its nascent form still requires the exploitation of animals. While a case can be made that even this is defensible if in vitro meat replaces conventional meat and the larger amount of harm it causes, our concern is not confined to in vitro meat in its current proto-typical form, and not only because producers are already trying to develop plant-based growth serum (Watson 2017). The deeper reason is that our inquiry is philosophical rather than historical. However else cultured meat may be important, and whether it takes two years or two hundred years for an entirely plant-based version to arrive, its central philosophical significance is that it provides an occasion to reconceive meat. We have long thought it an essential property of meat is that it is made from the body of a once-living animal. Considering cultured meat should spur us to reject this idea. In vitro meat is unique insofar as it raises the possibility of meat that does not require harming an animal and so is ethical to consume.

Philosophers have long drawn a distinction between a concept and a conception (Rawls 1971: 5). As Michael Hardimon puts it, "the concept of X provides the most basic possible characterization of X. A *conception* provides a particular way of articulating a concept" (2012: 258). If we were to outline the bare concept of justice, for example, we might characterize it as some kind of standard of fairness, perhaps one that should apply on a public level. *Conceptions* of justice do more. They go on to spell out the details of what a concept entails. Hence the familiar disagreement between conceptions of justice that favor or oppose the redistribution of wealth, legal access to abortion, affirmative action and other familiar flashpoints of controversy. As deep as the disagreements are between proponents of competing conceptions, each can still recognize the other as a (mistaken) conception of justice, rather than a conception of something else. The concept is necessarily shared across conceptions whereas the conceptions themselves are mutual rivals.

One way we sometimes come to see the difference between the concept and a conception of something is because the thing in question undergoes a material change. Consider, for example, pre-historic societies in which animals were used as currency. It is easy to imagine someone in such a society not giving much thought to the concept of currency. But suppose that coins come into circulation, and the pre-historic person suddenly realizes that animals are not part of currency's conceptual essence. Similarly, someone asked a hundred years ago to define the concept of a car could be forgiven for giving an answer that included reference to an internal combustion engine and a human driver. In the modern age of electric and self-driving cars, we recognize that gasoline and human direction are not essential to the concept of a car. Rather, they reflect a particular historical conception of cars. In a similar way, many societies have recently revised their marriage laws to include formations other than one man and one woman. In so doing they have recategorized heterosexual union, once thought to be essential to the concept of marriage, as one among multiple marriage conceptions.

Cultured meat is our occasion to view meat in a similar way. Being sourced in the body of an animal has long been thought to be part of the very concept of meat. In reality, it belongs only to a particular conception of meat, the traditional one that has long governed how we view it.

What *is* essential to the concept of meat? To answer this question it is helpful to distinguish between three aspects of meat: its origin, its substance, and its function (Sebo forthcoming: 13).[1]

Meat's traditional origin is easily identified. We have long taken it to necessarily come from a once-living animal. Meat's substance is something else. It is organic material made up of particular combinations of protein, water, and other building blocks. As for meat's function, it is on one level

[1] Corresponding to Aristotle's material, formal, and final causes, respectively.

phenomenological: it is something we eat with a particular taste and "mouthfeel," as food scientists put it. But meat's function is also nutritional: it is digestible and can be part of a healthy or unhealthy diet, depending on how we prepare it, how much we consume, and other considerations.

All three aspects of meat – origin, substance, and function – are referenced in the marketing material of Beyond Meat, a plant-based meat company:

> Meat is actually pretty simple: amino acids, fats, carbohydrates, trace minerals and water [substance] combined to give us that familiar chew, resistance and variation [function]. But what if we are able to take these same inputs from plants and combine them to look and feel just like animal meat [origin]? What you'd have is meat for the future. Meat from plants [origin] (Donaldson 2016: 195).

Beyond Meat is a company that specializes in plant-based meat, i.e., imitation hamburgers and chicken patties. Plant-based is not the same as cultured meat, the latter having a cellular rather than plant base. But this description of plant-based meat is worth noting because it advances a conception of meat that retains the traditional substance and function of meat but not its traditional origin. In vitro meat is similar insofar is it, too, retains meat's substance and function but not its traditional origin in the carcass of an animal.

There are precedents for food concepts that are defined by shared substance and function but not a shared origin. Cow's milk, coconut milk and mother's milk are all milks despite their different origins, coming from an animal, plant, and human being respectively. All three are naturally occurring light-colored liquids that we can safely drink and which have a taste that appeals to large numbers of people. While one might say the same of semen and perhaps other bodily fluids, milks are distinct from such fluids in the roles they play in reproduction and other bodily functions. Thus the concept of milk is broad enough to include white plant juices but not so broad as to contain every white mammalian bodily fluid. If a broad range of foods can share a concept in the case of milk, why can they not also do so in the case of meat?

One reason someone might reject this approach is that they sympathize with labeling systems that associate food products with geographic origin. France's *appellation d'origine contrôlée* (AOC) code, for example, famously precludes labeling as champagne any sparkling wine that is not produced in the Champagne region of France. Re-conceptualizing meat so as not to require an origin in an animal body might seem to threaten the rationale of the AOC and similar codes that grant weight to origin when determining whether sparkling wine or cheese warrant classification as "champagne" or "Parmigiano-Reggiano." It might seem a problem if my view were in conflict with a controversial but nevertheless widely accepted approach to categorization. One reason is because AOC approaches may capture a deep truth about food products. Alternatively, even if the AOC approach is wrong, its sheer popularity might be thought to pose a problem, insofar as people who accept it may be disinclined to accept my approach, however justified it may be.

Both problems are illusory. The approach to labeling represented by AOC certification is not based solely on origin. It is based on the view that origin affects the taste and other features of food products. As the USA Champagne Bureau puts it, "Location is particularly important to wine because the grapes and wines gain specific attributes due to the soil, air, water, and weather conditions of the place in which they are grown" (2017). On this view, geographic origin matters because it affects the substance and function of the food product. The reconceptualization I have put forward is not at odds with this approach to categorizing food products, because I do not take issue with the relevance of substance and function, only origin, when reconceptualizing meat.

An approach to conceptualizing food that would be at odds with mine is one that said origin of a food product is essential for its own sake. In the case of wine, this view would say the reason only sparkling wine from Champagne can be called Champagne is because of where it is from, period, and not because of any considerations related to the substance or function of wine, such as its taste. This view, however, is not worth taking seriously. An approach to food concepts that made origin, geographic or otherwise, essential for its own sake, even in cases in which substance and function remained the same, would be giving inappropriate emphasis to origin, emphasis that is at odds with how we normally think about food concepts such as milk.

On the view I am defending, what is essential to meat as a concept is its substance and function. Animal- and lab-sourced meat can thus both be genuine forms of meat as both are consistent with its concept. Eating cultured meat will therefore be consistent with animal protection insofar as it does not harm sentient animals. The same however is not true of animal-sourced meat. That practice remains unethical because of the harm done to the animal the meat comes from. If we were ever in a situation where the only choices were to eat animal- versus lab-sourced meat, it would be obligatory to eat the latter.

One possible objection to my argument might be that my revised concept of meat strays too far from what we have traditionally meant by "meat." If someone were to propose redefining the concept of an election so as to no longer denote a voting-based decision procedure, we would respond that rather than redefining the concept of an election, they have simply abandoned it. There is a limit on how far a concept can be redefined while still denoting something broadly similar to the original conceptualization. In such a spirit, someone might retort that it is an essential feature of meat that it be derived from the body of an animal. As my concept does not include this feature, it is not really a concept of meat.

Against this objection I would briefly make three points. This rival view amounts to the claim that something can be meat only if it is sourced in an animal, but this is implausible. Imagine the most perfect offering of in vitro meat conceivable. It flawlessly mimics traditional meat in every way save origin: in taste, nutrition, and composition. Were we to serve it to someone, they would have no way of knowing that it did not come from an animal. A meat concept that is not able to recognize meat that matches what we have traditionally meant by meat in all these ways is too narrow. If it looks like meat, cooks like meat and tastes like meat, then it is meat.

It also bears noting that my concept of meat still recognizes as meat the version that comes from animals. Rather than choose between the two, my argument has the function of widening the concept of meat so as to include both animal- and lab-sourced versions. There is thus no implication of no longer being able to use the term *meat* so as to apply to its traditional referents. Finally, the term *in vitro meat* is already in wide usage. This suggests the concept defender here, rather than being an eccentric revision of the concept of meat, may in fact accord with an already widespread but inchoate understanding. Indeed, one way to understand the concept of meat argued for here is an articulation of the notion of meat as we already encounter it in our period of history. As ever, the owl of Minerva spreads its wings only with the falling of the dusk.

The Edibility Objection

Although cultured meat has the potential to significantly reduce animal harm, it has faced a certain amount of resistance from animal protection theorists. (but see Singer [2013] for a protectionist proponent). Perhaps the most sweeping objection has been made by Rebekah Sinclair, one that if correct would render in vitro meat morally problematic for anyone to consume. Sinclair's argument as presented is directed at plant-based meat substitutes: vegetarian hot dogs and hamburgers that are widely found in supermarkets and marketed under brands such as Tofurky in North America or Fry's in the United Kingdom. If Sinclair's criticism of so-called meatless meats is sound, however, it will entail that in vitro meat is, if anything, *more* ethically suspect than its plant-based supermarket counterparts.

Sinclair argues that meatless meats further an ideology of animal edibility. To be edible in Sinclair's sense is to occupy a political rather than natural category: it is socially approved for such a being to be killed and eaten without risk of social or criminal sanction. Sinclair argues that the best response to the harm done to food animals is to reject outright the notion that they can be eaten. Not only do meatless meats fail to do this, she argues, but they "perpetuate a framework of intelligibility" according to which animals are edible (2016: 239). Meatless meats mimic traditional meat, and so depend on a pre-existing framework that sanctions eating animal bodies. This is evident in how imitation chicken and similar products are advertised: "the plant protein that looks, tastes, feels and acts like chicken—without the cluck" (236). Although such products are often put forth as a replacement for animal flesh, their substance and function are made intelligible by reference to the substance and function of animal-sourced meat. As a result, the challenge to meat eating such products represent is illusory, as they themselves presuppose eating animals as practice and ideology. "One cannot have meat without a death," Sinclair writes, "and it is this death that haunts the discourse of all meat (plant or otherwise)" (232).

Sinclair uses a vivid analogy to make her case. It involves a particular type of sex toy:

> Inflatable or silicone vaginas are a genre of sex toys which not only provide a vagina-shaped hole into which interested parties insert a sexual organ or object, but also (and quite often) come attached to entire sections of the female torso, buttocks and breasts. These are different than "fleshlights" or other male masturbators, which may or may not look like actual vaginas, and which do not reference the female anatomy in name or advertisement (2016: 237).

Sinclair's objection is not to the very idea that a device might provide sexual pleasure: sex toys as such are not her target. It is rather that one particular type of sex toy, silicone vaginas, are only understandable by reference to actual vaginas. Sinclair notes that artificial vaginas are sometimes advertised as being based on real models. They are "just like the real thing," or "lifelike," and potential users are invited to "feel what it's really like to be inside the world's hottest porn star" (237). As with meatless meat, we cannot understand these devices except by reference to the body parts they simulate. Sexual contact with a silicone vagina is unintelligible outside of the context of sex with an actual vagina. In this way the device relies on a pre-existing ideology that views female bodies as usable, even dismemberable, for the sake of providing pleasure to others. As with meatless meats, the object in question can only be understood and used within a framework that says the thing it imitates is "accessible and consumable" (239).

In the case of women, this framework divides the female body and detaches its parts from female agency, such that there is no requirement to obtain consent before engaging in sexual contact. According to Sinclair, silicone vaginas perpetuate rather than challenge the ideology of women's bodies being unconditionally available. In this way, silicone vaginas illustrate the same problematic substitution process of meatless meats. In both cases an artifact substitutes for something that is or was part of a living being, but the substitution is made possible by an ideology of objectification and consumption.

Sinclair's objection is total in that it suggests a moral wrong would occur if anyone consumed non-animal meats. However, several considerations suggest that this conclusion is too sweeping. I will shortly argue that consuming cultured meat (and by extension plant-based meat) is morally permissible, but before doing so, it is worth questioning what would actually follow from Sinclair's argument if it were sound. So let us grant for the sake of argument that meatless and in vitro meats are morally problematic in the way Sinclair suggests. Even so, it would still be permissible to promote in vitro meat's wide-scale adoption on the grounds that it was a lesser evil. I will focus on in vitro meat, but the argument will legitimize plant-based meat also.

Consider the following disturbing scenario. We are members of a humanitarian organization that has been granted access to a prisoner-of-war camp where some of the guards sexually assault the prisoners on a systematic basis. A defining feature of the scenario is that there is nothing we can do in the short term to make the guards stop viewing the prisoners as sexual objects. However, we learn an important piece of information about the guard-assailants. If they look in on the prisoners while they are undressing or sleeping naked this will satisfy their desire to objectify the prisoners and promptly end the assaults. Under such conditions it would be reasonable to promote the Peeping Tom option to the guards. This is consistent with viewing Peeping Tom behavior as morally wrong and otherwise to be discouraged and punished. Under the circumstances however, where it is a replacement for something even worse, it becomes acceptable to promote it as the less bad option.

A similar logic sees cultured meat as permissible to promote here and now even if it is also morally flawed. So long as it is replacing something worse, harming animals to make meat, it can be both morally compromised and something to be promoted. We can, of course, debate what alternative options might be available beyond cultured meat as a harm-reduction tool. My point, however, is not to suggest that encouraging the consumption of cultured meat is our only option. It is rather to illustrate the gap between being morally imperfect and not to be promoted. Therefore, even if everything Sinclair says it correct, it is not enough to discredit the view that the widespread adoption of in vitro meat is something to be wished for.

But *is* in vitro meat morally flawed? Sinclair's argument for this conclusion would seem to entail that many other objects that we normally consider morally acceptable are objectionable. Consider a child's doll or stuffed animal. Both are only intelligible by reference to the human being or animal they represent. There are many things we do to such toys, such as put words in their mouth and move them around as we please, that we would not consider acceptable in the case of the living beings they represent. Yet it does not seem morally problematic to treat inanimate objects in this way. This suggests that an artifact being intelligible by reference to a living thing is not problematic in and of itself. On this view, what Sinclair's analysis really suggests is problematic is an object being intelligible not by reference to a living being, but by reference to *mistreating* a living being.

Sinclair suggests something like this view when she explains why her objection is directed at only a subset of sex toys:

> Unlike inflatable and silicone vaginas, dildos are employable (and enjoyable) by men and women alike, and by both queer and straight bodies. This pivotal difference highlights a crucial element of the substitutionary structure we have been tracking. Namely, we are not only criticizing the process of dismemberment and substitution though that is obviously a central part of my critique; we are also tracking the way dominant carnophallogocentri[sm] . . . is secured through these substitutions and dismemberments. In other words, we are not

> only looking at the process of dismemberment as such, but also *who* is dismembered and *who* is substituted for *whose* sake (239).

This passage distinguishes artificial meats and artificial vaginas from sex toys that might also be thought to substitute for body parts, but which Sinclair's analysis does not rule out. The problem with silicone vaginas and plant-based meats is that they presume and legitimize "carnophallogocentricism," which denotes an ideology that, among other shortcomings, combines speciesism with sexism. Other toys, however, particularly those that can be used by people of diverse genders and sexualities, can be used in a manner that does not promote this ideology, and so do not represent the particular process of symbolic dismemberment Sinclair highlights. In short, pernicious substitution is a product not merely of how an item is designed but of how it is used.

If this is the case, however, it means Sinclair's criticisms will not apply to silicone vaginas and in vitro meats so long as they also can be used in a non-pernicious way. The question then becomes whether such uses are conceivable. In the case of highly realistic vaginas, we might imagine a heterosexual man who is disfigured or has some other condition that makes him unattractive to all of his potential sexual partners, yet who wants some sense of the experience of intercourse before he dies. Alternatively, we might imagine a woman who commissions the creation of a synthetic copy of her vagina. Suppose she were dying, and she and her partner had an especially intense commitment to monogamy. Under these conditions there might be meaning and value for him in having a means of engaging in symbolic intercourse with her even after her death. In terms of actually existing usages, Amazon currently markets "wearable vaginas" to transgender people and an article in a medical journal discusses the use of artificial vaginas in the study of ejaculatory dysfunction (Kobori et al. 2009). These and perhaps other uses would appear no more pernicious than those associated with other sex toys to which Sinclair is not opposed. But if so, then the issue is not with the very existence of silicone vaginas but how they are used.

The same is true of in vitro meat. One can imagine instances of consuming it that do not advance a pernicious ideology. To see this it is helpful to distinguish two ways in which someone might eat in vitro meat. On the one hand someone might eat it *and* animal-sourced meat. Alternatively, someone might eat only in vitro meat. The two meats are interchangeable in the first case but not the second. Sinclair's analysis blurs this important distinction. Someone who eats cultured meat while continuing to eat its animal-sourced counterpart clearly endorses the edibility of animals. While persuading this person to eat less animal-sourced and more in vitro meat can have value as a means of reducing animal harm, so far as the person's own ideology is concerned, it does indeed endorse traditional omnivorism.

But someone who eats *only* in vitro meat falls into a different category. Their diet is consistent with a doctrine of equal concern extended to animals. As such it rejects the edibility of animals in the clearest and most obvious way – by not eating them. In the case of this diet, synthetic meat is not merely an imitation but a *substitute*. The substitution is justified precisely by the fact that eating animals is morally wrong. And if the relevant question is for whose sake the substitution is made, then in this case it is done for the animals' sake. Insofar as doing so reduces the number of animals killed for food it will be incorrect to associate harm-reduced meats with an anti-animal ideology. Sexism is a cardinal example of a doctrine that fails to grant equal consideration to the interests of human beings. The same principle when extended to animals should urge us to welcome innovative solutions that have potential to make a positive difference for their treatment.

The equality-advancing use of in vitro meat should cause us to see that the analogy silicone vaginas is badly made. It would be well made if the facts of life were very different, such that every time a woman had vaginal sex she experienced some harm and synthetic vaginas were promoted as a means of harm reduction. Additionally, the analogy only works when we have in mind the use of an artificial body part as an imitation rather than a substitute. But synthetic vaginas are not put forward as something that should replace conventional vaginal sex. The strongest argument for in vitro meat, however, defends it as a substitution rather than a mere imitation.

But this is not the ultimate problem with Sinclair's argument. Her analysis is tied to a view of meat that views its animal origin as part of its very concept, hence her remark that one cannot have meat without a death. But this is precisely what in vitro meat calls into question. Once we realize this, we see that Sinclair's argument, although ostensibly radical, suffers from an excess of conservativism. It is conservative in the sense of falsely taking an aspect of the status quo as immutable. In this case, it is the traditional meaning of meat. But as I have argued, the correct response to in vitro meat is to regard it as occasion to refine our understanding of the concept of meat such that it does not essentially involve an animal origin.

Finally, Sinclair's argument suffers from a flawed understanding of intelligibility. She argues that because in vitro meat is intelligible by reference to old meat it perpetuates the ideology of traditional omnivorism. This overlooks how a concept can be intelligible by reference to a traditional meaning that it nonetheless escapes. There is a long history of terms that were once insults being embraced as positive labels. Quaker. Tory. Suffragette. Queer. The first time these terms were used in a positive way they were rendered intelligible by their history as an insult. But this avenue of intelligibility did nothing to change the fact that their new usage had a different meaning. This is possible because being intelligible by reference to a familiar meaning is not the same as being *defined* by that meaning. In the case of in vitro meat the meaning that is called into question is the concept of meat itself. It evokes the notion of meat as animal flesh precisely in order to overcome it.

The ultimate flaw of the intelligibility objection is that it is based on a false and self-defeating view of concepts that denies the possibility of new meanings springing up from the embers of the old. Certainly a positive feature of Sinclair's analysis is that in its own way it is motivated by a concern for animals. But taking that concern seriously means we should embrace in vitro meat rather than condemn it. Like Quakers and their counterparts, vegans see a concept they want to reclaim. In an age when animal protection is increasingly less marginal, meat is the first thing we should come for.

The Irreverence Objection

Imagine a detective, Clousseau, who investigates the home of an especially gruesome serial killer. The killer is in the habit of using human skin and bones to make lampshades, wastepaper baskets and other household objects. Suppose Clousseau is horrified by a human-skin lampshade he comes across in the killer's home. Even so, he cannot help but admire its aesthetic properties. He believes that taking the lampshade would disrespect the person whose skin it was made of and bestow a kind of retrospective validation on the killer's handiwork. So he does not want the actual human-skin lampshade. Instead he has a replica made, "from a synthetic material that closely simulates the feel and peculiar translucence of a middle-aged woman's skin. He puts it in his living room. He basks in its glow" (Fischer and Ozturk 2017: 1).

So write Bob Fischer and Burkay Ozturk, who devised the Clousseau thought experiment (2017). They argue that Clousseau commits a moral wrong. His lampshade is the product of a failure to exhibit the virtue of reverence. They cite a definition of reverence which characterizes it as "the well-developed capacity to have the feelings of awe, respect and shame" in circumstances in which these feelings are apt (1). Having a fake human lampshade is a failure to embody this virtue, one that represents "a kind of callousness, a thoughtlessness about how human bodies are symbols for the lives we live in them" (2).

Fischer and Ozturk suggest Clousseau exhibits a second moral failing. He admires something bad. As they explain, "After all, if Clousseau didn't admire Gein's handiwork, which is clearly bad, it's hard to see why Clousseau would have wanted a replica. This suggests that the replica might be a proxy for the real thing, so that enjoying the former is a way of enjoying the latter" (3).

Fischer and Ozturk draw an analogy between Clousseau's lampshade and fake meat. By fake meat they have in mind both in vitro meat as well as plant-based meats that are designed to imitate animal flesh to a significant degree. Two current examples are the so-called Beyond Burger and the Impossible Burger, which have attracted media attention for being plant-based burgers that mimic the properties of beef with a new degree of verisimilitude. Both products, for example, "bleed" in a manner reminiscent of traditional hamburgers, unlike previous vegetarian hamburgers. "Like Clousseau's replica, fake meat is supposed to resemble the real thing. Indeed, people consume fake meat precisely because it does resemble the real thing, and they value those particular gustatory qualities," Fischer and Ozturk write (3).

Fischer and Ozturk conclude that in vitro meat is morally problematic (they say the same of imitation-quality plant-based meat as well). To say it is problematic, however, is not to say no one should ever eat it. Indeed, they conclude that most people *should* in fact consume in vitro meat! The majority of the population is currently omnivorous. Insofar as cultured meat replaces animal-sourced meat, an important step forward will have occurred, for the wrong of eating in vitro meat is dwarfed by the wrong of animal meat, which harms sentient beings. There is no analogy to this in the original Clousseau thought experiment, but as the authors note, it can be revised to acknowledge this consideration. In the revised scenario, Clousseau loves the killer's original human-skin lampshade, and is tempted to make one of his own. Insofar as Clousseau's choice is between a real human lampshade and an imitation, he would be justified in creating the latter.

This justification, however, is not available to vegans or vegetarians. Insofar as they do not eat meat, there cannot appeal to harm reduction as a reason for eating cultured meat. In their case there is only the problematic nature of the product itself. As a result, vegans and vegetarians who eat cultured meat commit a moral wrong. A distinctive feature of Fischer and Ozturk's argument is thus that it offers separate conclusions regarding the permissibility of cultured meat for omnivores and vegans.

Addressing the Irreverence Objection

What should we make of the Clousseau thought experiment? We need to distinguish between two failings Fischer and Ozturk associate with Clousseau. One is his use of the synthetic-skin lampshade as a stand-in for the authentic skin lampshade. As the authors note, it is hard to see why Clousseau would have any interest in copying the killer's handiwork unless he admired it in some way. "This suggests that the replica might be a proxy for the real thing, so that enjoying the former is a way of enjoying the latter" (3). Call this the *enjoyment-by-proxy* problem. It is distinct from another problem, which asks how it is possible that Clousseau might admire the killer's lampshade in the first place despite his knowledge of how it was made. This would only be possible if Clousseau can "quarantine his knowledge of [the killer's] activities from his appreciation for the replica" (5). Call this the *quarantined knowledge* problem.

The enjoyment-by-proxy problem has no necessary analogue in the case of in vitro meat. If someone were to enjoy in vitro meat as a proxy for the enjoyment of animal flesh we could characterize their action as wrong because it reflects the wrong attitude – not because in vitro meat is wrong in itself. Imagine an alternative universe in which in vitro meat is invented but no one ever ate animals. It is just happenstance that the lab-created food product happens to mimic animal flesh. Of course such a world is improbable, but the point at hand is conceptual. Insofar as it is possible to imagine someone in such a world enjoying in vitro meat, this is enough to show that such enjoyment need not function as a proxy for enjoying animal flesh. It could be based simply on how in vitro meat tastes. And *this* ground for enjoyment is possible in any world, including ours. It is possible for someone to enjoy the taste of in vitro meat while not enjoying at all the thought of animals being killed for food. Once we recognize this, it should become clear that the Clousseau scenario stacks the deck against in vitro meat by building in an unnecessary association with the enjoyment of death or suffering.

In response to this criticism Fischer has replied that even if the association between in vitro meat and the enjoyment of animal harm is not necessary on a conceptual level, it is nonetheless a recurring association. It may be a contingent association, but in reality, it is a consistently recurring one. "In the actual world, animal products have that association, and so that's how we assess the virtuousness or viciousness of someone's attitudes."[2] This approach is reasonable, but insofar as it distinguishes in vitro meat from the attitude people adopt while consuming it, it concedes my point. While it is legitimate to criticize the attitudes people may exhibit while consuming in vitro meat or any other product, the issue at hand will be the attitudes themselves. Insofar as the food product in question does not necessarily give rise to any particular attitude, the product per se is not targeted by attitude-based criticism. In other words, if the ethics of eating in vitro meat are conditional on the attitude of the eater, then it can be permissible to eat it so long as it is done by someone who does not exhibit the problematic attitude.

This leaves the quarantined knowledge problem. A curious feature of how Fischer and Ozturk present it is that it characterized as a problem particular to vegetarians. As they put it, if it is disturbing that Clousseau can quarantine his knowledge, "then it should be disturbing that vegetarians can quarantine their knowledge of slaughter and dismemberment from their appreciation for fake meat" (5). By knowledge Fischer and Ozturk have in mind information about the suffering of food animals. Vegetarians, they write, have "seen videos of what happens in factory farm or read detailed descriptions of how animals are raised in CAFOs [concentrated animal feeding operations]" (5). The suffering of food animals, however, is not the only consideration that compels people to become vegetarian. Many arguments for not eating meat, including the one I have presented in this book, cite as an independent and sufficient reason the harm done to animals when they are killed. Very few people are unaware that meat is made from animals: such knowledge does not require research into animal husbandry practices. Insofar as there is a problem in quarantining one's knowledge of this point, it is a problem that can be exhibited by non-vegetarians also. Indeed, the everyday practice of meat-eating, which involves terms such as *beef* and *pork* that do not call to mind cows and pigs, as well as frequent efforts on the part of the meat industry to avoid scrutiny and publicity, may in fact depend on quarantining knowledge of the fate of food animals.

[2] Fischer, personal communication.

The quarantined knowledge problem, in short, presupposes a rationale for vegetarianism that appeals to animals suffering alone. But even if we grant for the sake of argument that reducing suffering is the only rationale for vegetarianism, omnivores would still not avoid the force of the problem. One reason is because some omnivores will also be aware of the suffering of food animals. Anyone who has seen a video of factory farms, let alone worked in one, will have access to the information Fischer and Ozturk emphasize. More generally, however, omnivores will face their own equivalent of the quarantine problem.

If quarantining one's knowledge of food animals suffering is wrong, it seems wrong because it involves not having animals suffering as part of one's top-of-mind awareness. The information in question is viewed as the kind of knowledge a person should keep in mind, to the point that they can be morally criticized for not doing so. By that standard, though, most omnivores who never even look into the plight of food animals also do wrong. In their case, they fail to live up to the obligation of having food animal suffering in mind, not by pushing it to the margins of their consciousness, but by never obtaining the information in the first place. It does not seem reasonable to criticize young children for not knowing about food animal suffering, and there may be rare cases of extenuating circumstances that excuse the ignorance of some adults. But for most members of modern society it is not difficult to come by knowledge of the suffering inflicted on food animals. If vegetarians can be criticized for coming across such information but then not paying it adequate mind, it seems only consistent to also criticize meat-eaters who fail to have such information in mind for a different reason: they never learned it in the first place even though they could have easily done so.

If that is the case, however, the quarantine problem does not give rise to a unique problem for vegetarians that renders in vitro meat impermissible for them alone. Rather, that same standard of epistemological expectation that generates the quarantine problem condemns omnivores and renders in vitro meat impermissible for them as well. It is easy to imagine pragmatic reasons why someone might want to hold omnivores and vegetarians or vegans to a different standard. But on a philosophical level there is no reason why the standard of expectation that gives rise to the quarantined-knowledge problem would discriminate according to diet. As such, the upshot of Fischer and Ozturk's analysis is broader than they suggest. "Some virtues – such as reverence – require *not* quarantining certain aspects of our mental lives from the rest" (6). By that standard, omnivores are also guilty of a failure of reverence.[3]

However, if we view in vitro meat the way I have characterized it, a reply is available to the quarantine problem. It is that being sourced in an animal's body is not an essential component of meat. We can disassociate in vitro meat from the process according to which animals are turned into food. When the concept of meat is correctly understood, killing sentient beings to use their bodies is not a necessary aspect. Insofar as the death of animals is not required, there *is* no knowledge of their deaths to quarantine. If this is so, the same reasoning will prevent the quarantine problem arising for plant-based mock meats, including not only bleeding veggie burgers but traditional vegetarian hot dogs, hamburgers and other meats. Indeed, the fact that these food items have long been marketed as hamburgers, hot dogs, etc. suggests that these and other meat-related concepts need not refer to animal flesh. In vitro meat correctly understood extends the same thinking to meat as such.

Here, then, are my responses to the enjoyment-by-proxy problem and the quarantined knowledge problem. In regards to the former, I have argued that in vitro meat need not be enjoyed as a proxy. In regards to the latter, I have argued that the concept of meat does not require animal flesh, so the quarantine problem simply does not arise. It bears noting, however, that even if I am wrong on both counts, a case can still be made for the universal permissibility of in vitro meat.

This can be seen by noting that even if we accept Fischer and Ozturk's thought experiment as stipulated, there is a slight disanalogy between Clousseau's lampshade and cultured meat. In the case of the inspector's replica, human flesh is given a new function. In our society it is not at all normal for lampshades to be made out of human skin. Clousseau's item, like that of the killer he models it on, employs skin in a new way. Cultured meat, by contrast, performs a function that was previously performed by animal-sourced meat. This function was entrenched at a society-wide level.

[3] Fischer and Ozturk discuss a case in which someone could be forgiven for quarantining their knowledge of animal suffering. "We might expect that a moral exemplar (someone who is virtuous, refrains from enjoying the bad, etc.) would make an effort to keep both associations intact unless there is a good reason (such as reduction or prevention of harms) to break it" (7). If I am correct that omnivores also fail to exhibit the virtue of reverence, they will not be moral exemplars on Fischer and Ozturk's account, and so this means of avoiding the quarantine problem will not be available to them. In a personal communication, Fischer writes, "quarantining knowledge is a bad-making feature of an attitude, but certainly one that can be outweighed. In the case of omnivores, it is indeed outweighed by the benefit of abstaining from eating real animal products." I argue below that an outweighing defense is also available to vegetarians and vegans.

In the human case there is a social narrative of decreasing respect that is not present in the animal case. Could this difference between the two replicas justify a different attitude toward the two items?

One way it can is if we view the transition from animal-sourced to in vitro sourced meat as part of a larger narrative of helping animals. This narrative does not end with the abolition of eating animals, radical as that would be in itself. Rather, it can be viewed as part of a larger change in how they are treated: as forms of entertainment, subjects of medical experiments, and all the familiar practices that animal protection opposes. The adoption of in vitro meat, even if it did exhibit a failure of reverence toward animals, is plausibly viewed as a step on the road to this outcome. This is not the same as saying that the failure of reverence that cultured meat is here taken to represent, and which I am here granting for the sake of argument, would necessarily be part of the final favored outcome. It could be a transitional step, like a ladder that we climb and then kick away. Cultured meat can be viewed as an instrument of transitional justice, whereas human-skin lampshades cannot.

This view is not necessarily opposed to Fischer and Ozturk's analysis. As we saw, they argued that cultured meat consumption is permissible for people who come to it as omnivores but not for non-meat eaters. But even in the latter case, they ask "How bad is it to eat fake meat? Perhaps not very." (4) If in vitro meat can be regarded as an instrument of transitional justice, however, this suggests that even if is wrong for vegans or anyone to eat cultured meat, it is a very small wrong, and one that is outweighed by seeking to contribute systematic social change. On this view, we would eat in vitro meat in order to encourage its widespread adoption, hoping to hasten the end of animal agriculture by providing an example of how it is possible to enjoy meat without harming animals.

Fischer and Ozturk conclude that the wrong of eating in vitro meat, while it may be small, is still serious enough to prevent vegetarians and vegans from doing so. One could accept their argument in its entirety, however, and view it as supporting a different conclusion. Rather than rule out in vitro meat, their argument instead entails modifying in vitro meat so that it does not exactly mimic animal meat. Both the substance and function could be changed. It would still function as a replacement, but it would no longer be an exact replica of traditional meat.

One way of doing so would be to making cultured meat taste slightly different, and perhaps better, than traditional meat. Perhaps someday it will be possible to do this while simultaneously making it more nutritious. Alternatively, the taste and nutritional content might stay the same, but the exact mix of protein and chemicals would change to be substantially non-identical to traditional meat. The first change would concern function, the second substance. Either way, in vitro meat 2.0, as we might dub it, would not be modelled on any existing meat but would represent a new variation. Currently, some existing breeds of cattle date to the ancient world while others were first bred only in the past two or three centuries. Modern breeds resemble older ones without being identical. Creating a new form of in vitro beef or other meat that did not precisely replicate the flesh of existing animal breeds would represent a change to the substance of such meats. At such point, with the origin, substance, and function all different, eating such a product would no longer seem to represent a failure of reverence for animals, as one was eating something that differed from their flesh at the level of origin, function, *and* substance. There would, again, be no failure of reverence.

At this point, an objection to my view is likely to arise. It is that it would seem to open the door to permitting synthetic human flesh. A central aspect of the Clousseau scenario, after all, is to make us grapple with the ethics of creating a facsimile of human flesh. If being sourced in the body of an animal is not necessary for meat, then by the same logic being sourced in the body of a human is not necessary to create human skin. So my argument might be thought to provide no grounds on which to condemn the creation of in vitro human flesh that was created to be eaten. What would we say about a Clousseau who used synthetic human parts not for a lampshade, but as a lamb chop?

The notion of synthetic human flesh or skin does not seem objectionable in itself. To see this, consider a revised version of the Clousseau scenario. Once again he creates synthetic human skin. In this case, however, rather than use it to shade his lamp, he provides it to surgeons performing skin graft operations. Clousseau is a great medical inventor and his synthetic skin is especially suited to this function, making grafts possible for patients who would never receive them otherwise. The function of Clousseau's synthetic skin in this scenario is far more positive than admiring the handiwork of a killer. As such, it does not seem right to associate our reconceived Clousseau with a failure of reverence for human lives and bodies, as his invention heals bodies in a manner that allows people's lives to go better. This suggests that if eating in vitro human flesh is objectionable, it is because it is being eaten, not the mere fact of in vitro human flesh as such (cf. Schaefer and Savulescu 2014).

As my view is a non-speciesist view, it offers no grounds to ascribe moral significance to species membership. As such, the fact that a piece of in vitro meat resembled human as opposed to non-human flesh would not be a morally relevant consideration. As such, there would be no grounds to treat the human kind different from the animal kind. If one can be eaten, both can. But also on my view, the human equivalent of in vitro meat would not really be human flesh. Just as meat no longer essentially involves the body of an animal, the concept of flesh no longer essentially would involve the body of a human being. Rather, we would distinguish human flesh from flesh as such.

Consuming in vitro flesh that tasted like human flesh would be morally impermissible insofar as it was guilty of the enjoyment-by-proxy failing. But insofar as someone was motivated to eat synthetic flesh for other reasons, no moral wrong would have occurred. Possible alternative motives might be cultural. The Wari' people of the western Amazonian rainforest are said to have once practiced a version of endocannibalism, the eating of members of one's own community. Anthropologists have used the term "compassionate cannibalism" to describe the custom: "funerary cannibalism was a main component of a grieving process in which causing the body of the deceased 'to disappear' helped the family and friends to come to terms with their loss more easily and more quickly" (Aubry 2002: 433). Additionally, some western women choose to eat their placenta after giving birth.[4] These practices illustrate the possibility of being motivated to eat human flesh for reasons other than a callous disregard for human life.[5] Unlike Smith's argument examined in the previous chapter, they do not entail that it is "ontologically illusory" to be a non-cannibal. As with in vitro meat, the ethics of consuming in vitro flesh will vary according to the motivation behind it. In neither case is it right to view producing the in vitro substance as wrong in itself.

My main aim has been to argue that animal protection theorists are wrong to reject in vitro meat. However, it is worth noting a further significant feature of cultured meat. If or when it becomes widespread, it has the potential to dissolve a traditional opposition between veganism and an opposing view that emphasizes food's status as a source of pleasure.

A hedonist approach to food is represented by celebrity chef Anthony Bourdain. "I look at food as something that should be pleasurable first. So the idea of creating synthetic meat, it seems antithetical to everything I believe in," Bourdain has said. "I see it as the enemy, any time you discuss

[4] This practice has been criticized for relying on false claims about the nutritional value of placenta (Sabin 2017).

[5] For further discussion of permissible instances of cannibalism see Milburn (2016).

food as sustenance first, without any consideration for happiness, joy, magic even" (Protin and Wei 2016). In vitro meat is the product of a framework concerned with the origin of meat, not, as Bourdain wrongly suggests, its sustenance. Despite this, what is worth noting about Bourdain's remark is its emphasis on a function of food: in this case, providing pleasure.

The current goal of in vitro meat production is to successfully imitate the texture and mouthfeel of animal-sourced meat. The focus on this goal is understandable given a desire to replace harm-sourced meat in the near-term. But mastery of the ability to create cultured meat holds out a further possibility. It could someday be *more* pleasurable than traditional meat. Animal bodies come with built in biological limits. This can render them recalcitrant and stubborn in face of the many things we seek to do to them. In vitro meat springs out of a mastery of the building blocks of flesh. It raises the possibility of mastery to the point of being able to manipulate the end product so as to please human taste buds even more than sirloin and pheasant. Make no mistake, Bourdain's appetite-driven view of animals is an example of what Bernard Williams called the "cruel superficiality of everyday thought" (1995: 218). But even when everyday hedonism is taken on its own terms, it is false to say that in vitro meat is antithetical to what it stands for. If hedonism and animal protection have traditionally had opposing approaches to food, this opposition is one that in vitro meat has the potential to someday overcome.

Bibliography

Abeyesinghe, S. M., C. J. Nicol, S. J. Hartnell, and C. M. Wathes. 2005. "Can Domestic Fowl, *Gallus Gallus Domesticus,* Show Self-Control?" *Animal Behaviour* 70 (1): 1–11.

Ainslie, George. 2001. *Breakdowns of Will.* Cambridge: Cambridge University Press.

Albers, Peter H., Gregory Linder, and James D. Nichols. 1990. "Effects of Tillage Practices and Carbofuran Exposure on Small Mammals." *The Journal of Wildlife Management* 54 (1): 135–42.

All Animals Australia. 2014. "Debunking 'Ordering the Vegetarian Meal? There's More Animal Blood on Your Hands.'" *All Animals Australia* (blog). November 12, 2014. http://allanimalsaustralia.wordpress.com/2014/11/12/debunking-ordering-the-vegetarian-meal-theres-more-animal-blood-on-your-hands.

Aloi, Giovanni. 2011. *Art and Animals.* London: I. B. Tauris.

Alpi, Amedio, Nikolaus Amrhein, Adam Bertl, Michael Blatt, Eduardo Blumwald, Felice Cervone, Jack Dainty, et al. 2007. "Plant Neurobiology: No Brain, No Gain?" *Trends in Plant Science* 12 (4): 135–6.

Alupay, Jean S., Stavros P. Hadjisolomou, and Robyn J. Crook. 2014. "Arm Injury Produces Long-Term Behavioral and Neural Hypersensitivity in Octopus." *Neuroscience Letters* 558 (January): 137–42.

Álvarez-Rivera, Manuel. 2018. "Election Resources on the Internet: Elections to the Dutch Tweede Kamer (House of Representatives) – Results Lookup." Electionresources.Org. http://electionresources.org/nl/house.php?election=2006.

American College of Obstetricians and Gynecologists. 2015. "How Your Fetus Grows During Pregnancy – ACOG." 2015. http://www.acog.org/Patients/FAQs/How-Your-Fetus-Grows-During-Pregnancy.

American Society for the Prevention of Cruelty to Animals. 2018. "Farm Animal Confinement Bans by State." ASPCA. 2018. http://www.aspca.org/animal-protection/public-policy/farm-animal-confinement-bans.

Anderson, Elizabeth. 2004. "Animal Rights and the Value of Non-Human Life." In *Animal Rights: Current Debates and New Directions*, edited by Cass R. Sunstein and Martha C. Nussbaum, 277–98. New York: Oxford University Press, USA.

Anderson, Roland C., Jennifer A. Mather, Mathieu Q. Monette, and Stephanie R. M. Zimsen. 2010. "Octopuses (*Enteroctopus Dofleini*) Recognize Individual Humans." *Journal of Applied Animal Welfare Science* 13 (3): 261–72.

Andre, Claire, and Manuel Velasquez. 1991. "Who Counts?" *Issues in Ethics* 4 (1): 1.

Animal Liberation South Australia. 2014a. "Archer's Dodgy Claims I . . . Health Advice." Animal Liberation South Australia. Accessed September 26, 2017. http://www.animalliberation.org.au/blog/2014/9/1/archers-dodgy-claims-i-health-advice.

2014b. "Archer's Dodgy Claims II . . . Land Use." Animal Liberation South Australia. Accessed September 26, 2017. http://www.animalliberation.org.au/blog/2014/9/1/archers-dodgy-claims-ii-land-use.

2014c. "Archer's Dodgy Claims III . . . Mouse Deaths." Animal Liberation South Australia. Accessed September 26, 2017. http://www.animalliberation.org.au/blog/2014/9/1/archers-dodgy-claims-iii-mouse-deaths.

Archer, Michael. 2011a. "Slaughter of the Singing Sentients: Measuring the Morality of Eating Red Meat." *Australian Zoologist* 35 (4): 979–82.

Archer, Michael. 2011b. "Ordering the Vegetarian Meal? There's More Animal Blood on Your Hands." *The Conversation.* http://theconversation.com/ordering-the-vegetarian-meal-theres-more-animal-blood-on-your-hands-4659.

American Dietetic Association. 2003. "Position of the American Dietetic Association and Dietitians of Canada: Vegetarian Diets." *Journal of the Academy of Nutrition and Dietetics* 103 (6): 748–65.

Aubry, Caroline. 2002. "Consuming Grief: Compassionate Cannibalism in an Amazonian Society." *Anthropological Quarterly* 75 (2): 433–6.

Austin, J. L. 1962. *Sense and Sensibilia.* Oxford: Oxford University Press.

Australian Export Grains Innovation Centre. 2016. "Australian Grain Production - a Snapshot." August 22, 2016. http://aegic.org.au/australian-grain-production-a-snapshot.

Australian Government: Department of Agriculture and Water Resources. 2017. "Wheat." Accessed September 25, 2017. http://www.agriculture.gov.au:80/ag-farm-food/crops/wheat.

Bailey, D. W., L. R. Rittenhouse, R. H. Hart, and R. W. Richards. 1989. "Characteristics of Spatial Memory in Cattle." *Applied Animal Behaviour Science* 23 (4): 331–40.

Bailey, Derek W., Larry D. Howery, and Darrin L. Boss. 2000. "Effects of Social Facilitation for Locating Feeding Sites by Cattle in an Eight-Arm Radial Maze." *Applied Animal Behaviour Science* 68 (2): 93–105.

Baillargeon, Renée, and Julie DeVos. 1991. "Object Permanence in Young Infants: Further Evidence." *Child Development* 62 (6): 1227–46.

Ball, Matt. 2005. "How Vegan?" *Vegan Outreach.* Accessed April 7, 2006. http://www.veganoutreach.org/enewsletter/20051005.html.

Barkham, Patrick. 2006. "Alfalfa Male Takes on the Corporation." *The Guardian*. September 8, 2006. http://www.theguardian.com/environment/2006/sep/08/food.ethicalliving.

Beiner, Ronald, and Jennifer Nedelsky. 2001. *Judgment, Imagination, and Politics: Themes from Kant and Arendt*. Rowman & Littlefield.

Bekoff, Marc. 2012. "Animals Are Conscious and Should Be Treated as Such." *New Scientist*. September 19, 2012. http://www.newscientist.com/article/mg21528836-200-animals-are-conscious-and-should-be-treated-as-such.

Benatar, David. 2006. *Better Never to Have Been: The Harm of Coming Into Existence*. New York: Oxford University Press.

Bennett, Jonathan. 1995. *The Act Itself*. Oxford: Oxford University Press.

Bermúdez, José Luis. 1996. "The Moral Significance of Birth." *Ethics* 106 (2): 378–403.

2013. "The Opacity of Mind: An Integrative Theory of Self-Knowledge, by Peter Carruthers." *Mind* 122 (485): 263–6.

Bernstein, Mark. 2004. *Without a Tear: Our Tragic Relationship with Animals*. Chicago: University of Illinois Press.

Block, Ned. 1981. "Psychologism and Behaviorism." *The Philosophical Review* 90 (1): 5–43.

Braithwaite, Victoria A., and Philip Boulcott. 2008. "Can Fish Suffer?" In *Fish Welfare*, edited by Edward Branson, 78–92. Oxford: Wiley-Blackwell.

Brenner, Eric D., Rainer Stahlberg, Stefano Mancuso, František Baluška, and Elizabeth Van Volkenburgh. 2007. "Response to Alpi et al.: Plant Neurobiology: The Gain Is More than the Name." *Trends in Plant Science* 12 (7): 285–6.

Brown, Culum, Kevin Laland, and Jens Krause. 2011. *Fish Cognition and Behavior*. Chichester, West Sussex: John Wiley & Sons.

Brown, Lester Russell. 2003. *Plan B: Rescuing a Planet Under Stress and a Civilization in Trouble*. New York: W. W. Norton & Company.

Bruckner, Donald. 2016. "Strict Vegetarianism Is Immoral." In *The Moral Complexities of Eating Meat*, edited by Ben Bramble and Bob Fischer, 30–47. New York: Oxford University Press, USA.

2015. "We Must Eat More Roadkill. Yes, Even You Vegetarians." *Salon*. November 28. http://www.salon.com/2015/11/28/we_must_eat_more_road kill_yes_even_you_vegetarians/.

Buchanan, Allen, Dan W. Brock, Norman Daniels, and Daniel Wikler. 2000. *From Chance to Choice: Genetics and Justice*. Cambridge: Cambridge University Press.

Budolfson, Mark Bryant. 2016a. "Consumer Ethics, Harm Footprints and the Empirical Dimensions of Food Choices." In *Philosophy Comes to Dinner: Arguments About the Ethics of Eating*, edited by Andrew Chignell, Terence Cuneo, and Matthew C. Halteman. New York: Routledge.

2016b. "Is It Wrong to Eat Meat from Factory Farms? If So, Why?" In *The Moral Complexities of Eating Meat*, edited by Ben Bramble and Bob Fischer. New York: Oxford University Press.

Bureau of Labor Statistics. 1999. "Industries with the Highest Nonfatal Total Cases, Incidence Rates for Injuries and Illnesses, Private Industry, 1999." Washington D.C.: U.S. Department of Labor.

Byrne, Alex. 2012. Review of *The Opacity of Mind: An Integrative Theory of Self-Knowledge,* by Peter Carruthers. *Notre Dame Philosophical Reviews*, May. http://ndpr.nd.edu/news/the-opacity-of-mind-an-integrative-theory-of-self-knowledge.

Cahoone, Lawrence. 2009. "Hunting as a Moral Good." *Environmental Values* 18 (1): 67–89.

Callicott, J. Baird. 2016. "The Environmental Omnivore's Dilemma." In *The Moral Complexities of Eating Meat,* edited by Ben Bramble and Bob Fischer. New York: Oxford University Press.

Carruthers, Peter. 1992. *The Animals Issue: Moral Theory in Practice*. Cambridge: Cambridge University Press.

2013. *The Opacity of Mind: An Integrative Theory of Self-Knowledge*. Oxford: Oxford University Press.

Case, David. 2010. "Investigation: Food That's Not Safe to Eat." *Animals in Print*. August 25, 2010. http://www.all-creatures.org/aip/nl-20100825–2.html.

Castrale, J. S. 1985. "Responses of Wildlife to Various Tillage Conditions." *Transactions of the North American Wildlife and Natural Resources Conference* (USA) 50: 142–56.

Caughley, Judy, Mary Bomford, Bob Parker, Ron Sinclair, John Griffiths, and Dana Kelly. 1998. *Managing Vertebrate Pests: Rodents*. Canberra: Bureau of Resource Sciences.

Cavia, R., I. E. G. Villafañe, E. A. Cittadino, D. N. Bilenca, M. H. Miño, & M. Busch, (2005). "Effects of Cereal Harvest on Abundance and Spatial Distribution of the Rodent Akodon azarae in Central Argentina." *Agriculture, Ecosystems & Environment*, 107 (1), 95–9.

Chambers, Philp, and Temple Grandin. 2004. *Guidelines for Humane Handling, Transport and Slaughtering of Livestock*. Bangkok: Food and Agriculture Organization of the United Nations.

Champaign Bureau, USA. 2017. "Protection of the Champagne Name." 2017. http://www.champagne.fr/en/comite-champagne/bureaus/bureaus/united-states/pages/protection-of-the-champagne-name.

Chandroo, K. P, I. J. H. Duncan, and R. D. Moccia. 2004. "Can Fish Suffer?: Perspectives on Sentience, Pain, Fear and Stress." *Applied Animal Behaviour Science*, International Society for Applied Ethology Special Issue: A selection of papers from the 36th ISAE International Congress., 86 (3): 225–50.

Chauhan, Chetan. 2013. "MoEF Panel Prohibits Dolphin Parks." *Hindustan Times*, January 8, 2013.

Churchill, Winston. 1931. "Fifty Years Hence." *The Strand*.

Clyne, Catherine. 2006. "Singer Says: The Satya Interview with Peter Singer." *Satya*. http://www.satyamag.com/octo6/singer.html.

Cochrane, Alasdair. 2012. *Animal Rights without Liberation: Applied Ethics and Human Obligations*. New York: Columbia University Press.

2013. "From Human Rights to Sentient Rights." *Critical Review of International Social and Political Philosophy* 16 (5): 655–75.

Cohen, Carl. 1986. "The Case for the Use of Animals in Biomedical Research." *New England Journal of Medicine* 315 (14): 865–70.

Connolly, Kate. 2002. "German Animals given Legal Rights." *The Guardian*, June 22, 2002, sec. World news.

Cook, Gareth. 2012. "Do Plants Think?" *Scientific American*. June 5, 2012. https://www.scientificamerican.com/article/do-plants-think-daniel-chamovitz.

Corliss, Richard. 2002. "Should We All Be Vegetarian?" *Time*, July 15, 2002.

Cowie, Christopher. 2017. "Does the Repugnant Conclusion Have Any Probative Force?" *Philosophical Studies* 174 (12): 3021–39.

Cozzutti, Claudio, and Giorgio Vallortigara. 2001. "Hemispheric Memories for the Content and Position of Food Caches in the Domestic Chick." *Behavioral Neuroscience* 115 (2): 305–13.

Dahlberg, Linda L., Robin M. Ikeda, and Marcie-jo Kresnow. 2004. "Guns in the Home and Risk of a Violent Death in the Home: Findings from a National Study." *American Journal of Epidemiology* 160 (10): 929–36.

Davidson, James N. 1997. *Courtesans & Fishcakes: The Consuming Passions of Classical Athens*. London: HarperCollins.

Davis, Karen. 2005. "UPC Reviews Temple Grandin's New Book, Animals in Translation." Upc-Online.Org. August 11, 2005. http://www.upc-online.org/whatsnew/81105grandin.htm.

Davis, Steven. 2003. "The Least Harm Principle May Require That Humans Consume a Diet Containing Large Herbivores, Not a Vegan Diet." *Journal of Agricultural and Environmental Ethics* 16 (4): 387–94.

DeGrazia, David. 1996. *Taking Animals Seriously: Mental Life and Moral Status*. Cambridge: Cambridge University Press.

1999. "Animal Ethics around the Turn of the Twenty-First Century." *Journal of Agricultural and Environmental Ethics* 11 (2): 111–29.

Demetriou, Dan, and Bob Fischer. 2018. "Dignitarian Hunting: A Rights-Based Defense." *Social Theory and Practice* 44 (1): 48–73.

Dennett, Daniel C. 1978. "Skinner Skinned." In *Brainstorms: Philosophical Essays on Mind and Psychology*, edited by Daniel C. Dennett, 53–70. Montgomery, Vt.: Bradford Books.

DeRamus, H. Alan, Terry C. Clement, Dean D. Giampola, and Peter C. Dickison. 2003. "Methane Emissions of Beef Cattle on Forages." *Journal of Environmental Quality* 32 (1): 269–277.

Dhar, Aarti. 2013. "India Bans Testing of Cosmetics on Animals." *The Hindu*, June 29, 2013,

Dombrowski, Daniel A. 1984. "Was Plato a Vegetarian?" *Apeiron* 18 (1): 1–9.

Donaldson, Brianne. 2016. "Exploiting Fantasy: Overconformity in Animal Agriculture, Meatless Meat and Animal Ethics." In *The Future of Meat without Animals*, edited by Brianne Donaldson and Christopher Carter. New York: Rowman and Littlefield.

Donaldson, Sue, and Will Kymlicka. 2011. *Zoopolis: A Political Theory of Animal Rights*. Oxford: Oxford University Press.

Doris, John Michael. 2015. *Talking to Our Selves: Reflection, Ignorance, and Agency*. Oxford: Oxford University Press.

Driver, Julia. 2016. "Individual Consumption and Moral Complicity." In *The Moral Complexities of Eating Meat*, edited by Ben Bramble and Bob Fischer, 67–79. New York: Oxford University Press.

Dworkin, Ronald. 1977. *Taking Rights Seriously*. Cambridge: Harvard University Press.

1983. "Comment on Narveson: In Defense of Equality." *Social Philosophy and Policy* 1 (1): 24–40.

2002. *Sovereign Virtue: The Theory and Practice of Equality*. Cambridge: Harvard University Press.

Edge, W. Daniel. 2001. "Wildlife of Agricultural, Pasture, and Mixed Environs." In *Wildlife-Habitat Relationships of Oregon and Washington*, edited by David H. Johnson and Thomas A. O'Neil, 342–60. Corvallis: Oregon State University Press.

Edge, W. Daniel, Jerry O. Wolff, and Robert L. Carey. 1995. "Density-Dependent Responses of Gray-Tailed Voles to Mowing." *The Journal of Wildlife Management* 59 (2): 245–51.

EFSA Scientific Committee. 2015. "Risk Profile Related to Production and Consumption of Insects as Food and Feed." *EFSA Journal* 13 (10): 1–60.

Eisnitz, Gail. 2006. "Why Not Change the Laws? Personal Communication." *Vegan Outreach*, January 24, 2006. http://www.veganoutreach.org/articles/whynotlaws.html#eisnitz.

Elvers, Barbara, ed. 2017. *Ullmann's Food and Feed*. Vol. 2. Weinheim: Wiley.

Emery, Nathan J., and Nicola S. Clayton. 2004. "Comparing the Complex Cognition of Birds and Primates." In *Comparative Vertebrate Cognition: Are Primates Superior to Non-Primates?*, edited by Lesley J. Rogers and Gisela Kaplan, 3–46. New York: Kluwer Academic.

Engel, Mylan, Jr. 2000. "The Immorality of Eating Meat." In *The Moral Life: An Introductory Reader in Ethics and Literature*, edited by Louis P. Pojman, 856–90. New York: Oxford University Press.

Ereshefsky, Marc. 2011. "Mystery of Mysteries: Darwin and the Species Problem." *Cladistics* 27 (1): 67–79.

European Union. 1999. "Council Directive 1999/74/EC." *Official Journal of the European Communities*, no. L 203 (July): 53–7.

2010. "Directive 2010/63/EU of the European Parliament and of the Council." *Official Journal of the European Union*, no. L 276 (September): 33–79.

Fairlie, Simon. 2010. *Meat: A Benign Extravagance*. White River Junction, Vt.: Chelsea Green Publishing.

FAOSTAT. 2011. "Statistical Database of the Food and Agricultural Organization of the United Nations." Food and Agricultural Organization of the United

Nations. Accessed October 15, 2011. http://faostat.fao.org/site/569/default.aspx#ancor.
Farm Sanctuary. 2005. "The Facts About Farm Animal Welfare Standards: A Summary of Farm Sanctuary's Research Report." Watkins Glen, NY: Farm Sanctuary.
Fegan, N., P. Vanderlinde, G. Higgs, and P. Desmarchelier. 2004. "The Prevalence and Concentration of Escherichia Coli O157 in Faeces of Cattle from Different Production Systems at Slaughter." *Journal of Applied Microbiology* 97 (2): 362–70.
Feltz, Adam, and Joshua May. 2017. "The Means/Side-Effect Distinction in Moral Cognition: A Meta-Analysis." *Cognition* 166 (1): 314–27.
Field, Tiffany M., Robert Woodson, Reena Greenberg, and Debra Cohen. 1982. "Discrimination and Imitation of Facial Expression by Neonates." *Science* 218 (4568): 179–81.
Firn, Richard. 2004. "Plant Intelligence: An Alternative Point of View." *Annals of Botany* 93 (4): 345–51.
Fischer, Bob. 2016. "Bugging the Strict Vegan." *Journal of Agricultural and Environmental Ethics* 29 (2): 255–63.
Fischer, Bob, and Andy Lamey. 2018. "Field Deaths in Plant Agriculture." *Journal of Agricultural and Environmental Ethics* 31 (4): 409–28
Fischer, Bob, and Burkay Ozturk. 2017. "Facsimiles of Flesh." *Journal of Applied Philosophy* 34 (4): 489–97.
Fitzpatrick, William J. 2006. "The Intend/Foresee Distinction and the Problem of 'Closeness'." *Philosophical Studies* 128 (3): 585–617.
Flegenheimer, Matt. 2013. "As Rats Escape Death, M.T.A. Turns to Sterilization." *The New York Times*, March 11.
Fleharty, Eugene D., and Kirk W. Navo. 1983. "Irrigated Cornfields as Habitat for Small Mammals in the Sandsage Prairie Region of Western Kansas." *Journal of Mammalogy* 64 (3): 367–79.
Forkman, Björn. 1998. "Hens Use Occlusion to Judge Depth in a Two-Dimensional Picture." *Perception* 27 (7): 861–7.
Foot, Philippa. 1967. "The Problem of Abortion and the Doctrine of Double Effect." *Oxford Review* 5: 5–15.
Fountain, Henry. 2013. "A Lab-Grown Burger Gets a Taste Test." *The New York Times*, August 5.
Fox, Michael Allen. 1999. *Deep Vegetarianism*. Philadelphia: Temple University Press.
Francione, Gary. 1996. *Rain without Thunder: The Ideology of the Animal Rights Movement*. Philadelphia: Temple University Press.
2000. *Introduction to Animal Rights: Your Child or the Dog?* Philadelphia: Temple University Press.
2008. *Animals as Persons: Essays on the Abolition of Animal Exploitation*. New York: Columbia University Press.
Franklin, Julian H. 2004. *Animal Rights and Moral Philosophy*. New York: Columbia University Press.

Frey, R. G. 1980. *Interests and Rights: The Case against Animals*. Oxford: Oxford University Press.

1983. *Rights, Killing and Suffering: Moral Vegetarianism and Applied Ethics*. Oxford: Blackwell.

Gagliano, Monica, Michael Renton, Martial Depczynski, and Stefano Mancuso. 2014. "Experience Teaches Plants to Learn Faster and Forget Slower in Environments Where It Matters." *Oecologia* 175 (1): 63–72.

Gallagher, Shaun. 1996. "The Moral Significance of Primitive Self-Consciousness: A Response to Bermúdez." *Ethics* 107 (1): 129–40.

Garnham, Laura, and Hanne Løvlie. 2018. "Sophisticated Fowl: The Complex Behaviour and Cognitive Skills of Chickens and Red Junglefowl." *Behavioral Science* 8 (1): 1–15.

Garrett, Aaron, ed. 2000. *Animal Rights and Souls in the 18th Century*. 6 vols. Bristol, UK and Sterling, Va.: Thoemmes Continuum.

Getz, Lowell L., and Elaine Brighty. 1986. "Potential Effects of Small Mammals in High-Intensity Agricultural Systems in East-Central Illinois, U.S.A." *Agriculture, Ecosystems & Environment* 15 (1): 39–50.

Gianessi, Leonard. 2014. "Importance of Herbicides for No-Till Agriculture in South America." Washington D.C.: Croplife Foundation. http://croplife.org/case-study/importance-of-herbicides-for-no-till-agriculture-in-south-america/.

Godfrey-Smith, Peter. 2013. "Cephalopods and the Evolution of the Mind." *Pacific Conservation Biology* 19 (1): 4–9.

2016. *Other Minds: The Octopus and the Evolution of Intelligent Life*. New York: HarperCollins.

Government Accountability Office. 2004. "Humane Methods of Slaughter Act: USDA Has Addressed Some Problems but Still Faces Enforcement." Washington D.C.

Grandin, Temple. 1980a. "Livestock Behavior as Related to Handling Facilities Design." *International Journal for the Study of Animal Problems* 1 (1): 33–52.

1980b. "Bruises and Carcass Damage." *International Journal for the Study of Animal Problems* 1 (1): 121–37.

1995. "Animal Behavior and the Design of Livestock and Poultry Systems." In *Proceedings from the Animal Behavior and the Design of Livestock and Poultry Systems International Conference*. Indianapolis, Indiana.

1998. "Review: Reducing Handling Stress Improves Both Productivity and Welfare." *The Professional Animal Scientist* 14 (1): 1–10.

2000. "Bruise Levels on Fed and Non-Fed Cattle." In *Proceedings Livestock Conservation Institute*, 193–201. Kansas City, Missouri.

2001. "Progress in Livestock Handling and Slaughter Techniques in the United States, 1970–2000." In *The State of the Animals 2001*, edited by A. Rowan and D Salem. Washington D.C.: Humane Society Press.

2006a. *Thinking in Pictures: And Other Reports from My Life with Autism*. Second ed. London: Bloomsbury.

2006b. "Progress and Challenges in Animal Handling and Slaughter in the U.S." *Applied Animal Behaviour Science* 100 (1): 129–39.

2009. "Poultry Slaughter Plant and Farm Audit: Critical Control Points for Bird Welfare (updated July 2009)." http://www.grandin.com/poultry.audit.html.

2010a. "Improving Livestock, Poultry and Fish Welfare in Slaughter Plants with Auditing Programs." In *Improving Animal Welfare: A Practical Approach*, edited by Temple Grandin, 160–85. Cambridge, MA: CABI.

2010b. "The Effect of Economic Factors on the Welfare of Livestock and Poultry." In *Improving Animal Welfare: A Practical Approach*, edited by Temple Grandin. Cambridge, MA: CABI.

2011. "Directions for Laying out Curved Cattle Handling Facilities for Ranches, Feedlots, and Properties." 2011. http://www.grandin.com/design/curved.handling.facilities.html.

2012. "Center Track Conveyor Restrainer for Beef Cattle." 2012. http://www.grandin.com/restrain/new.conv.rest.html.

2013. "The Effect of Economics on the Welfare of Cattle, Pigs, Sheep, and Poultry." 2013. http://www.grandin.com/welfare/economic.effects.welfare.html.

2014. "Recommended Captive Bolt Stunning Techniques for Cattle (updated October 2014)." 2014. http://www.grandin.com/humane/cap.bolt.tips.html.

2017. "Biography: Temple Grandin. PhD." Grandin.Com. http://www.grandin.com/temple.html.

Grandin, Temple, and Catherine Johnson. 2010. *Animals Make Us Human: Creating the Best Life for Animals*. Boston: Houghton Mifflin Harcourt.

Grandin, Temple, and Joe M. Regenstein. 1994. "Religious Slaughter and Animal Welfare: A Discussion for Meat Scientists." *Meat Focus International* 3 (1): 115–23.

Guzy, Carol. 2008. "Dawn in Louisiana." *The Washington Post*, August 15, 2008.

Hall, Matthew. 2011. *Plants as Persons: A Philosophical Botany*. Albany: SUNY Press.

Handfield, Toby. 2011. "Absent Desires." *Utilitas* 23 (4): 402–27.

Hardimon, Michael O. 2012. "The Idea of a Scientific Concept of Race." *Journal of Philosophical Research* 37 (April): 249–82.

Hare, R. M. 1972. "Rules of War and Moral Reasoning." *Philosophy & Public Affairs* 1 (2): 166–81.

Hare, R. M. 1999. "Why I Am Only a Demi-Vegetarian." In *Singer and His Critics*, edited by Dale Jamieson, 233–46. Oxford: Blackwell.

Hart, H. L. A. 1980. "Death and Utility." *The New York Review of Books*, May 15, 1980.

Hayden, Benjamin Y. 2016. "Time Discounting and Time Preference in Animals: A Critical Review." *Psychonomic Bulletin & Review* 23 (1): 39–53.

Henry, Ben Andrew. 2017. "Can Plants Learn to Associate Stimuli with Reward?" *The Scientist*. http://www.the-scientist.com/?articles.view/articleNo/48028/title/Can-Plants-Learn-to-Associate-Stimuli-with-Reward-/.

Heyd, David. 1994. *Genethics: Moral Issues in the Creation of People*. Berkeley: University of California Press.

Hodson, Gordon, and Megan Earle. 2018. "Conservatism Predicts Lapses from Vegetarian/Vegan Diets to Meat Consumption (through Lower Social Justice Concerns and Social Support)." *Appetite* 120 (January): 75–81.

Hooley, Dan, and Nathan Nobis. 2016. "A Moral Argument for Veganism." In *Philosophy Comes to Dinner: Arguments About The Ethics of Eating*. New York: Routledge.

Horta, Oscar. 2014. "The Scope of the Argument from Species Overlap." *Journal of Applied Philosophy* 31 (2): 142–54.

Humane Society of the United States. 2008. "An HSUS Report: The Welfare of Animals in the Egg Industry." *IIA*, January. http://animalstudiesrepository.org/hsus_reps_impacts_on_animals/16.

Hvorecny, Lauren M., Jessica L. Grudowski, Carrie J. Blakeslee, Tiffany L. Simmons, Paula R. Roy, Jennifer A. Brooks, Rachel M. Hanner, et al. 2007. "Octopuses (*Octopus Bimaculoides*) and Cuttlefishes (*Sepia Pharaonis, S. Officinalis*) Can Conditionally Discriminate." *Animal Cognition* 10 (4): 449–59.

IDAUSA. 2010. "Cosmetic Testing: Facts." In *Defense of Animals*. 2010. http://www.idausa.org/facts/costesting.html.

Irish Independent. 2009. "Eating Meat with an Easy Conscience (Editorial)." *Irish Independent*, January 13, 2009.

Jacob, Jens, and Nadine Hempel. 2003. "Effects of Farming Practices on Spatial Behaviour of Common Voles." *Journal of Ethology* 21 (1): 45–50.

Jacobs, Suzanne. 2015. "With Lab-Grown Meat, Can We Have Our Animals and Eat Them Too?" *Grist*. August 3, 2015. http://grist.org/food/with-lab-grown-meat-can-have-our-animals-and-eat-them-too/.

Jamieson, Dale. 1990. "Rights, Justice, and Duties to Provde Assistance: A Critique of Regan's Theory of Rights." *Ethics* 100 (2): 349–62.

Jenni, Kathie. 1994. "Dilemmas in Social Philosophy: Abortion and Animal Rights." *Social Theory and Practice* 20 (1): 59–83.

Jensen, Rikke A., Mary S. Wisz, and Jesper Madsen. 2008. "Prioritizing Refuge Sites for Migratory Geese to Alleviate Conflicts with Agriculture." *Biological Conservation* 141 (7): 1806–18.

Johnson, D. R. 1987. "Effect of Alternative Tillage Systems on Rodent Density in the Palouse Region." *Northwest Science* 61 (1): 37–40.

Johnston, Amy N., Thomas H. Burne, and Steven P. Rose. 1998. "Observation Learning in Day-Old Chicks Using a One-Trial Passive Avoidance Learning Paradigm." *Animal Behaviour* 56 (6): 1347–53.

Jones, Dana. 2008. "Crimes without Consequences: The Enforcement of Humane Slaughter Laws in the United States." Washington D.C.: Animal Welfare Institute.

Kazez, Jean. 2017. "The Taste Question in Animal Ethics." *Journal of Applied Philosophy*: 1–14.

Keller, Simon. 2005. "Comments on George Schedler, 'Should Peter Singer Become an Ethical Meat Eater?'" *Southwest Philosophy Review* 21 (2): 159–62.

Kertész, Á., and B. Madarász. 2014. "Conservation Agriculture in Europe." International Soil and Water Conservation Research 2 (1): 91–6.

Kobori, Yoshitomo, Ryo Sato, Yoshio Ashizawa, Hiroshi Yagi, Shigehiro So, Gaku Arai, and Hiroshi Okada. 2009. "Rehabilitation for Ejaculatory Dysfunction Using a Masturbator." *The Japanese Journal of Impotence Research* 24 (3): 355–7.

Korsgaard, Christine M. 2005. *Fellow Creatures: Kantian Ethics and Our Duties to Animals*, edited by G. Peterson. Vol. 25/26. The Tanner Lecture on Human Values. Salt Lake City: University of Utah Press.

Kriegel, Uriah. 2013. "Animal Rights: A Non-Consequentialist Approach." In *Animal Minds and Animal Ethics*, edited by K. Petrus and M. Wild. Bielefeld, Germany: Transcript.

Kuba, Michael, Ruth Byrne, Daniela Meisel, and Jennifer Mather. 2006. "When Do Octopuses Play? Effects of Repeated Testing, Object Type, Age, and Food Deprivation on Object Play in *Octopus Vulgaris*." *Journal of Comparative Psychology* 120 (3): 184–90.

Kymlicka, Will. 1990a. *Contemporary Political Philosophy: An Introduction*. Oxford: Oxford University Press.

1990b. "Two Theories of Justice." *Inquiry* 33 (1): 99–119.

1991. "The Ethics of Inarticulacy." *Inquiry* 34 (2): 155–82.

Laland, Keven N., Culum Brown, and Jens Krause. 2003. "Learning in Fishes: From Three-Second Memory to Culture." *Fish and Fisheries* 4 (3): 199–202.

Lamey, Andy. 2007. "Food Fight! Davis Versus Regan on the Ethics of Eating Beef." *Journal of Social Philosophy* 38 (2): 331–48.

2011. "The Ethical Creature: Animals and Equal Consideration." PhD Thesis, Perth, WA: University of Western Australia.

2017. "Review of *The Moral Complexities of Eating Meat*." *Between the Species* 20 (1): 133–44.

Langley, Ricky Lee, and James Lee Hunter. 2001. "Occupational Fatalities Due to Animal-Related Events." *Wilderness & Environmental Medicine* 12 (3): 168–74.

Lea, Stephen E. G., Alan M. Slater, and Catriona M. E. Ryan. 1996. "Perception of Object Unity in Chicks: A Comparison with the Human Infant." *Infant Behavior and Development* 19 (4): 501–4.

Leahy, Michael P. T. 1991. *Against Liberation: Putting Animals in Perspective*. New York: Routledge.

Leblanc-Fournier, Nathalie, Ludovic Martin, Catherine Lenne, and Mélanie Decourteix. 2014. "To Respond or Not to Respond, the Recurring Question in Plant Mechanosensitivity." *Frontiers in Plant Science* 5: 401.

Lee, Susan J., Henry J. Peter Ralston, Eleanor A. Drey, John Colin Partridge, and Mark A. Rosen. 2005. "Fetal Pain: A Systematic Multidisciplinary Review of the Evidence." *JAMA* 294 (8): 947–54.

Levy, Neil. 2004. "Cohen and Kinds: A Response to Nathan Nobis." *Journal of Applied Philosophy* 21 (2): 213–7.

2016. "Vegetarianism: Toward Ideological Impurity." In *The Moral Complexities of Eating Meat*, edited by Ben Bramble and Bob Fischer, 172–84. New York: Oxford University Press.

Littlejohn, Clayton. 2016. "Potency and Permissibility." In *The Moral Complexities of Eating Meat*, edited by Ben Bramble and Bob Fischer. New York: Oxford University Press.

Low, Philip, Jaak Panksepp, Diana Reiss, David Edelman, Van Swideren, and Christof Kock. 2012. "The Cambridge Declaration on Consciousness."

Luper, Steven. 2009. *The Philosophy of Death*. Cambridge: Cambridge University Press.

Lynch, J. 2001. "NW's Biggest Slaughterhouse under Fire for Tactics." *The Oregonian*, January 24, 2001.

Lynch, Rene. 2008. "'Hell's Kitchen': Bye-Bye Ben." *Los Angeles Times*. May 13, 2008. http://latimesblogs.latimes.com/showtracker/2008/05/hells-kitchen-b.html.

MacDonald, Fiona. 2015. "This Indoor Farm Is 100 Times More Productive than Outdoor Fields." ScienceAlert. http://www.sciencealert.com/this-indoor-farm-is-100-times-more-productive-than-an-outdoor-one.

MacIntyre, Alasdair. 1981. *After Virtue*. Notre Dame: University of Notre Dame Press.

Mangels, Ann Reed, and Suzanne Havala. 1994. "Vegan Diets for Women, Infants, and Children." *Journal of Agricultural and Environmental Ethics* 7 (1): 111–22.

Marcelis, L. F. M., and S. Hemming. 2013. "Greenhouse Production in The Netherlands." *Resource Magazine* 20 (2): 14.

Marcus, Erik. 2005. *Meat Market: Animals, Ethics, and Money*. Ithica: Brio Press.

Marder, Michael. 2012a. "If Peas Can Talk, Should We Eat Them?" *The New York Times*, April 28.

2012b. "Is Plant Liberation on the Menu?" *The New York Times*, May 8.

2012c. "Plant Intentionality and the Phenomenological Framework of Plant Intelligence." *Plant Signaling & Behavior* 7 (11): 1365–72.

2013. *Plant-Thinking: A Philosophy of Vegetal Life*. New York: Columbia University Press.

Marino, Lori. 2017. "Thinking Chickens: A Review of Cognition, Emotion, and Behavior in the Domestic Chicken." *Animal Cognition* 20 (2): 127–47.

Marquis, Don. 1989. "Why Abortion Is Immoral." *The Journal of Philosophy* 86 (4): 183–202.
Matheny, Gaverick. 2003. "Least Harm: A Defense of Vegetarianism from Steven Davis's Omnivorous Proposal." *Journal of Agricultural and Environmental Ethics* 16 (5): 505–11.
Matheny, Gaverick, and Kai M. A. Chan. 2005. "Human Diets and Animal Welfare: The Illogic of the Larder." *Journal of Agricultural and Environmental Ethics* 18 (6): 579–94.
Mather, Jennifer A. 2008. "Cephalopod Consciousness: Behavioural Evidence." *Consciousness and Cognition* 17 (1): 37–48.
Matthews, Joseph. 1957. *Reporting Wars*. Oxford: Oxford University Press.
May, Joshua. 2014. "Moral Judgment and Deontology: Empirical Developments." *Philosophy Compass* 9 (11): 745–55.
2018. *Regard for Reason in the Moral Mind*. Oxford: Oxford University Press.
McDonald's. 2018. "McDonald's Commitment to Chicken Sustainability." *News*. mcdonalds.com. March 26, 2018. http://news.mcdonalds.com/stories/using-our-scale-for-good-details/commitment-chicken-sustainability?cid=PS:GCM_BR:B::Google:mcdonalds_animal_welfare.
McLeod, Ross. 2004. Counting the Cost: Impact of Invasive Animals in Australia, 2004. Canberra: Cooperative Research Centre for Pest Animal Control.
McMahan, Jeff. 2002. *The Ethics of Killing: Problems at the Margins of Life*. New York: Oxford University Press.
2008. "Eating Animals the Nice Way." *Daedalus* 137 (1): 66–76.
2009. "Asymmetries in the Morality of Causing People to Exist." In *Harming Future Persons: Ethics, Genetics and the Nonidentity Problem*, edited by David Wasserman and Melinda Roberts, 49–68. Dordrecht: Springer.
2016a. "The Moral Problem of Predation." In *Philosophy Comes to Dinner: Arguments About the Ethics of Eating. London: Routledge*, edited by Andrew Chignell, Terence Cuneo, and Matthew C. Halteman, 268–94. New York: Routledge.
2016b. "The Comparative Badness for Animals of Suffering and Death." In *The Ethics of Killing Animals*, edited by Tatjana Višak and Robert Garner. Oxford: Oxford University Press.
2017. "Might We Benefit Animals by Eating Them?" Youtube. LSE Philosophy. June 5, 2017. http://www.youtube.com/watch?v=2tZ–bCOUro.
McPherson, Tristram. 2016. "Why I Am a Vegan (And You Should Be One Too)." In *Philosophy Comes to Dinner: Arguments About the Ethics of Eating*. New York: Routledge.
Meltzoff, Andrew N., and M. Keith Moore. 1977. "Imitation of Facial and Manual Gestures by Human Neonates." *Science* 198 (4312): 75–78.
1983. "Newborn Infants Imitate Adult Facial Gestures." *Child Development* 53 (4): 702 9.
Metheringham, Jill. 2000. "Disposal of Day-Old Chicks: The Way Forward." *World Poultry* 16 (11): 25, 27.

Meyers, C. D. 2013. "Why It Is Morally Good to Eat (Certain Kinds of) Meat: The Case for Entomophagy." *Southwest Philosophy Review* 29 (1): 119–26.

Milburn, Josh. 2016. "Chewing Over In Vitro Meat: Animal Ethics, Cannibalism and Social Progress." *Res Publica* 22 (3): 249–65.

Molteni, Megan. 2017. "NYC's Newest Weapon Against the Rats? Sterilization." *Wired*. http://www.wired.com/2017/04/nycs-newest-weapon-rats-sterilization.

Moe, Randi Oppermann, Janicke Nordgreen, Andrew M. Janczak, Berry M. Spruijt, Adroaldo J. Zanella, and Morten Bakken. 2009. "Trace Classical Conditioning as an Approach to the Study of Reward-Related Behaviour in Laying Hens: A Methodological Study." *Applied Animal Behaviour Science* 121 (3): 171–78.

Nagel, Thomas. 1974. "What Is It Like to Be a Bat?" *The Philosophical Review* 83 (4): 435–50.

1979. *Mortal Questions*. Cambridge: Cambridge University Press.

1991. *Equality and Partiality*. Oxford: Oxford University Press.

Nahmias, Eddy. 2002. "When Consciousness Matters: A Critical Review of Daniel Wegner's The Illusion of Conscious Will." *Philosophical Psychology* 15 (4): 527–41.

2014. "Is Free Will an Illusion? Confronting Challenges From the Modern Mind Sciences." In *Moral Psychology, Vol. 4: Freedom and Responsibility*, edited by Walter Sinnott-Armstrong. Cambridge, Mass.: MIT Press.

Nass, Roger D., Glenn A. Hood, and Gerald D. Lindsey. 1971. "Fate of Polynesian Rats in Hawaiian Sugarcane Fields during Harvest." *The Journal of Wildlife Management* 35 (2): 353–56.

National Institute for Occupational Safety and Health. 1993. "Injuries among Farm Workers in the United States, 1993." http://www.cdc.gov/niosh/97–115.html.

Nelkin, Dana Kay, and Samuel Rickless. 2014. "Three Cheers for Double Effect." *Philosophy and Phenomenological Research* 89 (1): 125–58.

2015. "So Close, Yet So Far: Why Solutions to the Closeness Problem for the Doctrine of Double Effect Fall Short." *Noûs* 49 (2): 376–409.

New South Wales Government: Department of Primary Industries. 2017. "Mouse Biology." http://www.dpi.nsw.gov.au/biosecurity/vertebrate-pests/pest-animals-in-nsw/mouse-and-mice-plagues/mouse-biology.

Nicol, C. J., and S. J. Pope. 1992. "Effects of Social Learning on the Acquisition of Discriminatory Keypecking in Hens." *Bulletin of the Psychonomic Society* 30 (4): 293–96.

1994. "Social Learning in Small Flocks of Laying Hens." *Animal Behaviour* 47 (6): 1289–96.

1996. "The Maternal Feeding Display of Domestic Hens Is Sensitive to Perceived Chick Error." *Animal Behaviour* 52 (4): 767–74.

1999. "The Effects of Demonstrator Social Status and Prior Foraging Success on Social Learning in Laying Hens." *Animal Behaviour* 57 (1): 163–71.

Nilsson, Lovisa, Nils Bunnefeld, Jens Persson, and Johan Månsson. 2016. "Large Grazing Birds and Agriculture—Predicting Field Use of Common Cranes and Implications for Crop Damage Prevention." *Agriculture, Ecosystems & Environment* 219 (Supplement C): 163–70.

Nobis, Nathan. 2004. "Carl Cohen's 'Kind' Arguments for Animal Rights and against Human Rights." *Journal of Applied Philosophy* 21 (1): 43–59.

2016. "What's Wrong With Linking Abortion and Animal Rights?" What's Wrong? (blog). July 16, 2016. http://whatswrongcvsp.com/2016/07/16/whats-wrong-with-linking-abortion-and-animal-rights.

Norcross, Alastair. 2004. "Puppies, Pigs, and People: Eating Meat and Marginal Cases." *Philosophical Perspectives* 18 (1): 229–45.

Nordgreen, Janicke, Joseph P. Garner, Andrew Michael Janczak, Birgit Ranheim, William M. Muir, and Tor Einar Horsberg. 2009. "Thermonociception in Fish: Effects of Two Different Doses of Morphine on Thermal Threshold and Post-Test Behaviour in Goldfish (Carassius Auratus)." *Applied Animal Behaviour Science* 119 (1): 101–7.

Nozick, Robert. 1974. *Anarchy, State, and Utopia*. New York: Basic Books.

Nussbaum, Martha. 1998. "Still Worthy of Praise." *Harvard Law Review* 111 (7): 1776–95.

2007. *Frontiers of Justice: Disability, Nationality, Species Membership*. Harvard University Press.

Ober, Josiah. 2011. *Political Dissent in Democratic Athens: Intellectual Critics of Popular Rule*. Princeton: Princeton University Press.

O'Connor, Denis R. 2002. "Report of the Walkerton Inquiry: The Events of May 2000 and Related Issues." Toronto: Ontario Ministry of the Attorney General.

Otsuka, Michael. 2008. "Double Effect, Triple Effect and the Trolley Problem: Squaring the Circle in Looping Cases." *Utilitas* 20 (1): 92–110.

Parfit, Derek. 1984. *Reasons and Persons*. Oxford: Oxford University Press.

Pazol, Karen, Andreea Creanga, and Denise Jamieson. 2015. "Abortion Surveillance – United States, 2012." Centers for Disease Control and Prevention. Accessed May 12, 2018. http://www.cdc.gov/mmwr/preview/mmwrhtml/ss6410a1.htm.

Pech, R., G. Hood, G. Singleton, E. Salmon, R. Forrester, and P. Brown. 1999. "Models for Predicting Plagues of House Mice (*Mus domesticus*) in Australia." In *Ecologically-Based Management of Rodent Pests*. Canberra: Australian Centre for International Agricultural Research.

Pelizzon, Alessandro, and Monica Gagliano. 2015. "The Sentience of Plants: Animal Rights and Rights of Nature Intersecting." *Australian Animal Protection Law Journal* 11: 5–13.

Pincock, Stephen. 2007. "Meat, in Vitro?" *The Scientist*. Accessed August 24, 2018. http://www.the-scientist.com/notebook-old/meat-in-vitro-46164.

Pinkert, M., J. Meerbeek, G. Scholten, & J. Jenks. 2002. "Impact of Crop Harvest on Small Mammal Populations in Brookings County, South Dakota." *Proceedings of the South Dakota Academy of Science* 81: 39–45.
Plato. 1976. *The Republic*. Translated by A.D. Lindsay. London: J. M. Dent.
Pollan, Michael. 2002. "An Animal's Place." *The New York Times Magazine*.
2013. "The Intelligent Plant." *The New Yorker*. http://www.newyorker.com/magazine/2013/12/23/the-intelligent-plant.
Porcher, Jocelyne. 2017. *The Ethics of Animal Labor: A Collaborative Utopia*. Cham, Switzerland: Palgrave Macmillan.
Portmore, Douglas W. 2007. "Consequentializing Moral Theories." *Pacific Philosophical Quarterly* 88 (1): 39–73.
Posner, Richard. 2004. "Animal Rights: Legal, Philosophical and Pragmatic Perspectives." In *Animal Rights: Current Debates and New Directions*, edited by Cass Sunstein and Martha Nussbaum. New York: Oxford University Press.
Prince-Hughes, Dawn. 2004. *Songs of the Gorilla Nation: My Journey through Autism*. New York: Three Rivers.
Protin, Corey, and Will Wei. 2016. "Anthony Bourdain Slams Fake Meat: 'I See It as the Enemy.'" Business Insider. http://www.businessinsider.com/anthony-bourdain-big-problem-synthetic-fake-meat-laboratory-2016–12.
Quinn, Warren S. 1989. "Actions, Intentions, and Consequences: The Doctrine of Double Effect." *Philosophy and Public Affairs* 18 (4): 334–51.
Rachels, James. 1990. *Created from Animals: The Moral Implications of Darwinism*. Oxford: Oxford University Press.
Rawls, John. 1971. *A Theory of Justice*. Cambridge: Harvard University Press.
Raz, Joseph. 1986. *The Morality of Freedom*. Oxford: Oxford University Press.
Regan, Tom. 2004. *The Case for Animal Rights*. Second ed. Berkeley: University of California Press.
Regolin, Lucia, Rosa Rugani, Piero Pagni, and Giorgio Vallortigara. 2005. "Delayed Search for Social and Nonsocial Goals by Young Domestic Chicks, Gallus Gallus Domesticus." *Animal Behaviour* 70 (4): 855–64.
Regolin, Lucia, Giorgio Vallortigara, and Mario Zanforlin. 1994. "Perceptual and Motivational Aspects of Detour Behaviour in Young Chicks." *Animal Behaviour* 47 (1): 123–131.
1995. "Detour Behaviour in the Domestic Chick: Searching for a Disappearing Prey or a Disappearing Social Partner." *Animal Behaviour* 50 (1): 203–211.
Rogers, Lesley J., and Gisela Kaplan. 2004. "Preface." In *Comparative Vertebrate Cognition: Are Primates Superior to Non-Primates?*, edited by Lesley J. Rogers and Gisela Kaplan. New York: Kluwer Academic.
Rogers, Simon. 2010. "Maternal Mortality: How Many Women Die in Childbirth in Your Country?" *The Guardian*. April 13, 2010.
Rothgerber, Hank. 2015. "Can You Have Your Meat and Eat It Too? Conscientious Omnivores, Vegetarians, and Adherence to Diet." *Appetite* 84 (1): 196–203.

Sabin, Dyani. 2017. "'Don't Eat the Placenta,' Doctors Warn New Parents." *Livescience*. 2017. http://www.livescience.com/60680-eating-the-placenta-health-risks.html.

Salt, Henry. 1914. "The Logic of the Larder." In *The Humanities of Diet*. Manchester: The Vegetarian Society.

Salva, Orsola Rosa, Jonathan Niall Daisley, Lucia Regolin, and Giorgio Vallortigara. 2009. "Lateralization of Social Learning in the Domestic Chick, Gallus Gallus Domesticus: Learning to Avoid." *Animal Behaviour* 78 (4): 847–856.

Sapontzis, S. F. 1987. *Morals, Reason, and* Animals. Philadelphia: Temple University Press.

Schaefer, G. Owen, and Julian Savulescu. 2014. "The Ethics of Producing In Vitro Meat." *Journal of Applied Philosophy* 31 (2): 188–202.

Schedler, George. 2005. "Does Ethical Meat Eating Maximize Utility?" *Social Theory and Practice* 31 (4): 499–511.

Schlosser, Eric. 2002. *Fast Food Nation: The Dark Side of the All-American Meal*. New York: HarperCollins.

Schmidtz, David. 1998. "Are All Species Equal?" *Journal of Applied Philosophy* 15 (1): 57–67.

Schwer, Laura. 2011. "Small Mammal Populations in Switchgrass Stands Managed for Biomass Production Compared to Hay and Corn Fields in Kentucky." M.S. Thesis, Lexington: University of Kentucky.

Scruton, Roger. 2000. *Animal Rights and Wrongs*. Third ed. London: Metro.

Scully, Matthew. 2002. *Dominion: The Power of Man, the Suffering of Animals and the Call to Mercy*. New York: St. Martin's.

Searle, John R. 1980. "Minds, Brains, and Programs." *Behavioral and Brain Sciences* 3 (3): 417–24.

Sebo, Jeff. Forthcoming. "The Ethics and Politics of Plant-Based and Cultured Meat." *Les Ateliers de l'éthique/The Ethics Forum*.

Sher, George. 2014. *Equality for Inegalitarians*. Cambridge: Cambridge University Press.

Sherwin, C. M., C. M. Heyes, and C. J. Nicol. 2002. "Social Learning Influences the Preferences of Domestic Hens for Novel Food." *Animal Behaviour* 63 (5): 933–42.

Shoemaker, Natalie. 2015. "Price of Lab-Grown Burger Falls from $325 K to $11.36." Big Think. October 29, 2015. http://bigthink.com/ideafeed/answering-how-a-sausage-gets-made-will-be-more-complicated-in-2020.

Shoemaker, Sydney. 1994. "The First-Person Perspective." *Proceedings and Addresses of the American Philosophical Association* 68 (2): 7–22.

Shreeve, James. 2003. "Oliver's Travels." *The Atlantic*, October.

Shriver, Adam. 2009. "Knocking Out Pain in Livestock: Can Technology Succeed Where Morality Has Stalled?" *Neuroethics* 2 (3): 115–24.

2010. "Not Grass-Fed, but at Least Pain-Free." *The New York Times*, February 18.

Sinclair, Rebekah. 2016. "The Sexual Politics of Meatless Meat." In *The Future of Meat without Animals*, edited by Brianne Donaldson and Christopher Carter, 229–48. New York: Rowman and Littlefield.

Singer, Peter. 1975. *Animal Liberation: A New Ethics for Our Treatment of Animals.* New York: The New York Review of Books.

1979. *Practical Ethics.* Cambridge: Cambridge University Press.

1980. "Do Consequences Count? Rethinking the Doctrine of Double Effect." *Hastings Center Report* 10 (1): 42–44.

1985a. "Ten Years of Animal Liberation." *The New York Review of Books*, 1985.

1985b. "The Dog in the Lifeboat: An Exchange." *The New York Review of Books*, 1985.

1990. *Animal Liberation.* Second ed. New York: The New York Review of Books.

1993. *Practical Ethics.* Second ed. Cambridge: Cambridge University Press.

1998. "Possible Preferences." In *Preferences*, edited by Christophe Fehige and Ulla Wessels. 383–98. Berlin: Walter de Gruyter.

1999a. "A Response." In *Singer and His Critics*, edited by Dale Jamieson, 269–335. Oxford: Blackwell.

2000. *Ethics Into Action: Henry Spira and the Animal Rights Movement.* Lanham, Maryland: Rowman & Littlefield.

2011. *Practical Ethics.* Third ed. Cambridge: Cambridge University Press.

2013. "The world's first cruelty free hamburger." *The Guardian*, August 5, 2013.

2015. "The Abuse of Animals Won't Stop until We Stop Eating Meat." *The Guardian*, February 11, 2015.

2016. "Afterword." In *The Ethics of Killing Animals*, edited by Tatjana Višak and Robert Garner. 229–36. Oxford: Oxford University Press.

Singleton, Grant R., Peter R. Brown, Roger P. Pech, Jens Jacob, Greg J. Mutze, and Charles J. Krebs. 2005. "One Hundred Years of Eruptions of House Mice in Australia – a Natural Biological Curio." *Biological Journal of the Linnean Society* 84 (3): 617–27.

Skelhorn, John. 2011. "Colour Biases Are a Question of Conspecifics' Taste." *Animal Behaviour* 81 (4): 825–9.

Smart Floating Farms. 2017. "Smart Floating Farms." http://smartfloatingfarms.com/.

Smith, Andrew F. 2016. *A Critique of the Moral Defense of Vegetarianism.* Houndmills, Basingstoke: Palgrave Macmillan.

Sneddon, Lynne. 2007. "Assessing Pain Perception in Fish from Physiology to Behaviour." In *Comparative Biochemistry and Physiology Part A: Molecular & Integrative Physiology*, Abstracts of the Annual Main Meeting of the Society for Experimental Biology, Glasgow, Scotland, 31st March–4th April, 2007, 146 (4, Supplement): S78.

Sneddon, Lynne, and Matthew Leach. 2016. "Anthropomorphic Denial of Fish Pain." *Animal Sentience* 1 (3). http://animalstudiesrepository.org/animsent/vol1/iss3/28.

Sneddon, Lynne U. 2003. "The Evidence for Pain in Fish: The Use of Morphine as an Analgesic." *Applied Animal Behaviour Science* 83 (2): 153–62.

Solomon, Molly. 2016. "The Final Days Of Hawaiian Sugar." NPR.Org. Accessed September 26, 2017. http://www.npr.org/sections/thesalt/2016/12/17/505861855/the-final-days-of-hawaiian-sugar

Sorabji, Richard. 1995. *Animal Minds and Human Morals: The Origins of the Western Debate*. Ithaca: Cornell University Press.

Stenhouse, David. 1974. *Evolution of Intelligence*. London: Allen & Unwin.

Stephen, Leslie. 1896. *Social Rights and Duties: Addresses to Ethical Societies, Volume One*. London: S. Sonnenschein & Company.

Strasser, Annie-Rose. 2014. "The Weird And Wonderful World of Indoor Farming." *ThinkProgress*. Accessed September 26, 2017. http://thinkprogress.org/the-weird-and-wonderful-world-of-indoor-farming-333b6a12cdec/.

Struik, Paul C, Xinyou Yin, and Holger Meinke. 2008. "Plant Neurobiology and Green Plant Intelligence: Science, Metaphors and Nonsense." *Journal of the Science of Food and Agriculture* 88 (3): 363–70.

Stump, Scott. 2014. "TODAY Puts 'meatless' Meat to the Test: Does It Taste like Chicken?" TODAY.Com. April 25, 2014. http://www.today.com/food/today-puts-meatless-meat-test-does-it-taste-chicken-1D79579619.

Sunstein, Cass R. 2001. "Slaughterhouse Jive." *The New Republic,* January 28, 40–5

Tännsjö, Tobjörn. 2002. "Why We Ought to Accept the Repugnant Conclusion." *Utilitas* 14 (3): 339.

Taylor, Charles. 1989. *Sources of the Self: The Making of the Modern Identity*. Cambridge: Harvard University Press.

Taylor, Pollyanna E., Marie Haskell, Michael C. Appleby, and Natalie K. Waran. 2002. "Perception of Time Duration by Domestic Hens." *Applied Animal Behaviour Science* 76 (1): 41–51.

Taylor, Rowan. 2001. "A Step at a Time: New Zealand's Progress towards Hominid Rights." *Animal Law Review* 7: 35–43.

Tew, T. E., and D. W. Macdonald. 1993. "The Effects of Harvest on Arable Wood Mice Apodemus Sylvaticus." *Biological Conservation* 65 (3): 279–83.

Texas Parks and Wildlife. 2013. "Texas Hunting Incidents Analysis 2013." *Austin*. http://tpwd.texas.gov/publications/pwdpubs/media/pwd_rp_k0700_1124_2013.pdf.

The Avian Brain Nomenclature Consortium. 2005. "Avian Brains and a New Understanding of Vertebrate Brain Evolution." *Nature Reviews Neuroscience* 6 (2): 151–9.

Thomson, Judith Jarvis. 1985. "The Trolley Problem." *The Yale Law Journal* 94 (6): 1395–415.

1990. *The Realm of Rights*. Cambridge: Harvard University Press.

Tooley, Michael. 1972. "Abortion and Infanticide." *Philosophy & Public Affairs* 2 (1): 37–65.

1983. *Abortion and Infanticide*. Oxford: Oxford University Press.

Tooley, Michael, Peter Singer, and Helga Kluse. 2009. "Personhood." In *A Companion to Bioethics*, Second ed. Oxford: Wiley-Blackwell.

Trewavas, Anthony. 2007. "Response to Alpi et al.: Plant Neurobiology – All Metaphors Have Value." *Trends in Plant Science* 12 (6): 231–33.
United States Department of Agriculture. 2005. "U.S. Sugar Beet Farming–How 'Sweet' Is It?" United States Department of Agriculture. 2005. http://www.ers.usda.gov/AmberWaves/February05/Findings/Sugarbeets.htm.
2016. "Overview of the United States Slaughter Industry." United States Department of Agriculture. http://usda.mannlib.cornell.edu/usda/current/SlauOverview/SlauOverview-10–27-2016.pdf.
2017. "Livestock Slaughter 2016 Summary." United States Department of Agriculture. http://usda.mannlib.cornell.edu/usda/current/LiveSlauSu/LiveSlauSu-04–19-2017.pdf.
U.S. Environmental Protection Agency. 2006a. "Frequently Asked Questions." U.S. Environmental Protection Agency. http://www.epa.gov/methane/rlep/faq.html.
2006b. "U.S. Greenhouse Gas Inventory Reports Inventory of U.S. Greenhouse Gas Emissions and Sinks: 1990–2004." USEPA #430-R-06–002. April.
Vallortigara, Giorgio. 2004. "Visual Cognition and Representation in Birds and Primates." In *Comparative Vertebrate Cognition: Are Primates Superior to Non-Primates?*, edited by Lesley J. Rogers and Gisela Kaplan. 57–85. New York: Kluwer Academic.
Vallortigara, Giorgio, Lucia Regolin, Margherita Rigoni, and Mario Zanforlin. 1998. "Delayed Search for a Concealed Imprinted Object in the Domestic Chick." *Animal Cognition* 1 (1): 17–24.
Varner, Gary. 1994. "What's Wrong with Animal *by-Products*?" *Journal of Agricultural and Environmental Ethics* 7 (1): 7–17.
1998. *In Nature's Interests?: Interests, Animal Rights, and Environmental Ethics.* Oxford: Oxford University Press.
2012. *Personhood, Ethics, and Animals: Situating Animals in Hare's Two Level Utilitarianism.* Oxford: Oxford University Press.
Vegan Outreach. 2005. "Defining Vegan." *Vegan Outreach.* Accessed April 7, 2006. http://www.veganoutreach.org/advocacy/definingvegan.html.
Vegetarian Resource Group. 2016. "How Many Adults in the U.S. Are Vegetarian and Vegan?" Vegetarian Resource Group. 2016. http://www.vrg.org/nutshell/Polls/2016_adults_veg.htm.
Velleman, J. David. 1991. "Well-Being and Time." *Pacific Philosophical Quarterly* 72 (1): 48–77.
2000. "Well-Being and Time." In *The Possibility of Practical Reason*, 56–84. New York: Oxford University Press.
Višak, Tatjana. 2013. *Killing Happy Animals: Explorations in Utilitarian Ethics.* New York: Springer.

2016. "Do Utilitarians Need to Accept the Replaceability Argument?" In *The Ethics of Killing Animals*, edited by Tatjana Višak and Robert Garner. Oxford: Oxford University Press.

Walzer, Michael. 1987. *Interpretation and Social Criticism*. Cambridge, Mass.: Harvard University Press.

2006. *Just and Unjust Wars: A Moral Argument with Historical Illustrations.* New York: Basic Books.

Wang, Frances Kai-Hwa. 2016. "Hawaii's Last Sugar Plantation Finishes Its Final Harvest." NBC News. Accessed September 26, 2017. http://www.nbcnews.com/news/asian-america/hawaii-s-last-sugar-plantation-finishes-its-final-harvest-n695396.

Warburton, D. B., and W. D. Klimstra. 1984. "Wildlife Use of No-till and Conventionally Tilled Corn Fields." *Journal of Soil and Water Conservation* 39 (5): 327–30.

Warrick, Jo. 2001. "'They Die Piece by Piece.'" *Washington Post*, April 10.

Washington State Department of Agriculture. 2001. "Investigations into Alleged Violations of State Law by IBP Inc. Summary Report." Olympia, WA.

Watson, Elaine. 2017. "Hampton Creek to Enter Clean Meat Market in 2018: 'We Are Building a Multi-Species, Multi-Product Platform.'" Foodnavigator-Usa.Com. June 26, 2017. http://www.foodnavigator-usa.com/Article/2017/06/27/Hampton-Creek-to-enter-clean-meat-market-in-2018.

Wayne, Katherine. 2013. "Permissible Use and Interdependence: Against Principled Veganism." *Journal of Applied Philosophy* 30 (2): 160–75.

Wegner, Daniel M. 2002. *The Illusion of Conscious Will.* Cambridge, Mass.: MIT Press.

Wegner, John, and Gray Merriam. 1990. "Use of Spatial Elements in a Farmland Mosaic by a Woodland Rodent." *Biological Conservation* 54 (3): 263–76.

Weissmann, A., S. Reitemeier, A. Hahn, J. Gottschalk, and A. Einspanier. 2013. "Sexing Domestic Chicken before Hatch: A New Method for in Ovo Gender Identification." *Theriogenology* 80 (3): 199–205.

White, Alan. 1989. "Why Animals Cannot Have Rights." In *Animal Rights and Human Obligations*, edited by Tom Regan and Peter Singer, Second ed. Englewood Cliffs, New Jersey: Prentice Hall.

White, J., K. Hoskins, & J. Wilson (1998). "The Control of Rodent Damage in Australian Macadamia Orchards by Manipulation of Adjacent Non-Crop Habitats." *Crop Protection* 17 (4): 353–7.

Williams, Bernard. 1976. "Persons, Character and Morality." In *The Identities of Persons*, edited by Amélie Rorty, 197–216. Berkeley: University of California Press.

1995. "Replies." In *World, Mind, and Ethics: Essays on the Ethical Philosophy of Bernard Williams*, edited by J. E. J. Altham and Ross Harrison. Cambridge: Cambridge University Press.

Wisconsin Department of Natural Resources. 2006. "Cryptosporidium: A Risk to Our Drinking Water." http://dnr.wi.gov/org/water/dwg/crypto.Htm.

Wolf, Clark. 2004. "O Repugnance, Where Is Thy Sting?" In *The Repugnant Conclusion*, edited by Tobjörn Tännsjö and Jesper Ryberg, 61–80. Library Of Ethics And Applied Philosophy. Dordrecht: Springer.

Young, R. 1984. "Response of Small Mammals to No-Till Agriculture in Southwestern Iowa." M.S. Thesis, Ames: University of Iowa.

Zimmerman, P. H., S. A. F. Buijs, J. E. Bolhuis, and L. J. Keeling. 2011. "Behaviour of Domestic Fowl in Anticipation of Positive and Negative Stimuli." *Animal Behaviour* 81 (3): 569–77.

Index